CMOS Active Inductors and Transformers

Principle, Implementation, and Applications

Fei Yuan

CMOS Active Inductors and Transformers

Principle, Implementation, and Applications

 Springer

Fei Yuan
Department of Electrical and Computer Engineering
Ryerson University
Toronto, Ontario, Canada

ISBN 978-0-387-76477-1   e-ISBN 978-0-387-76479-5

Library of Congress Control Number: 2008925091

Printed on acid-free paper.

9 8 7 6 5 4 3 2 1

springer.com

This book is dedicated to
Jing

Preface

CMOS spiral inductors have found a broad range of applications in high-speed analog signal processing and data communications. These applications include bandwidth enhancement, delay reduction, impedance matching, frequency selection, distributed amplifiers, RF phase shifters, low-noise amplifiers, and voltage-controlled oscillators, to name a few. The effectiveness of these inductors, however, is affected by a number of limitations intrinsic to the spiral layout of the inductors. These limitations include a low quality factor, a low self-resonant frequency, a small and non-tunable inductance, and the need for a prohibitively large silicon area. The use of CMOS spiral transformers in RF applications such as low-noise amplifiers, power amplifiers, and LC oscillators has emerged recently. These transformers are constructed by coupling two spiral inductors via a magnetic link. They offer the advantages of a reduced silicon consumption and increased inductances. The limitations of spiral inductors, however, are inherited by spiral transformers.

Inductors and transformers synthesized using active devices, known as active inductors and transformers, offer a number of unique advantages over their spiral counterparts including virtually no chip area requirement, large and tunable inductances with large inductance tuning ranges, large and tunable quality factors, high self-resonant frequencies, and full compatibility with digital oriented CMOS technologies. Active inductors and transformers have found increasing applications in high-speed analog signal processing and data communications where spiral inductors and transformers are usually employed. As compared with spiral inductors and transformers, the applications of CMOS active inductors and transformers are affected by a number of limitations intrinsic to synthesized devices. These limitations include a small dynamic range, poor noise performance, a high level of power consumption, and a high sensitivity to supply voltage fluctuation and process variation.

This book provides a comprehensive treatment of the principle, topologies, and characteristics of CMOS active inductors and transformers, and an in-depth examination of their emerging applications in high-speed analog signal processing and data communications. The materials presented in the book are based on the work of many researchers who contributed to the theory and design of CMOS active inductors and transformers. In recognition of their contributions, the active inductors and transformers presented in this text are named in the names of the researchers. For active inductors and transformers developed by more than two researchers, although due to the space constraint, only the name of the first author of the work is used to name the active inductors and transformers, the contributions of all other authors are equally recognized. This is reflected by the presentation of the full authorship of the work in the *References* of the book. The same approach is followed in the presentation of CMOS active inductor/transformer bandpass filters, oscillators, and other sub-systems.

This book consists of two parts : Part I - *Principle and Implementation of CMOS Active Inductors and Transformers*, and Part II -*Applications of CMOS Active Inductors and Transformers*.

Part I of the book deals with the topologies, characteristics and implementation of CMOS active inductors and transformers. This part consists of three chapters.

Chapter 1 starts with a brief investigation into why inductive characteristics are critically needed in high-speed applications. This is demonstrated with the applications of inductors and transformers in LC oscillators, impedance matching networks, RF phase shifters, RF power dividers, frequency selection networks, in particular, RF bandpass filters, and low-noise amplifiers. A detailed examination of the design constraints of monolithic inductors and transformers is followed. The advantages and design challenges of CMOS active inductors and transformers are examined in detail.

Chapter 2 presents the principles of the synthesis of inductors using gyrator-C networks. Both lossless and lossy single-ended and fully differential gyrator-C active inductors are studied. The important figure-of-merits that quantify the performance of active inductors including frequency operation range, inductance tunability, quality factor, noise, linearity, stability, supply voltage sensitivity, parameter sensitivity, signal sensitivity, and power consumption are examined in detailed. The details of the CMOS implementation and analysis of single-ended and fully differential active inductors are presented. The circuit implementation and characteristics of published CMOS active inductors are examined in detail.

Chapter 3 focuses on the principles of the synthesis of CMOS active transformers. Both lossless and lossy gyrator-C active transformers are studied. The characterization of active transformers including stability, frequency operation range, the tunability of self and mutual inductances, turn ratios, coupling factors, voltage and current transfer characteristics, impedance transformation, noise, quality factors, linearity, supply voltage sensitivity, parameter sensitivity, and power consumption is examined in detail. The CMOS implementation of several published CMOS active transformers is presented and their characteristics are analyzed.

Part II of the book focuses upon the emerging applications of CMOS active inductors and transformers in high-speed analog signal processing and data communications. This part consists of four chapters.

Chapter 4 investigates the implementation and characteristics of RF bandpass filters using CMOS active inductors. The chapter starts with a detailed investigation of the characterization of bandpass filters. Bandwidth, 1-dB compression points, third-order intercept points, noise figure, noise bandwidth, spurious-free-dynamic range, frequency selectivity, and passband center frequency tuning are examined. It is followed by a detailed examination of the configurations of RF bandpass filters with active inductors. Wu bandpass filters, Thanachayanont bandpass filters, Xiao-Schaumann bandpass filters, Thanachayanont-Payne bandpass filter, and Weng-Kuo bandpass filters are studied and their performance is compared.

Chapter 5 looks into the realization of the building blocks of high-speed transceivers using CMOS active inductors and transformers. The use of CMOS active inductors in low-noise amplifiers, optical front-ends, RF phase shifters, RF modulators, RF power dividers, and Gb/s serial-link transceivers is examined in detail.

Chapter 6 starts with a brief review of the fundamentals of electrical oscillators. Both ring and LC oscillators are investigated. The use of CMOS active inductors in improving the performance of ring oscillators is investigated. The presentation continues with a close examination of the use of CMOS active inductors in LC oscillators. A special attention is given to the comparison of the phase noise of these oscillators. LC oscillators and LC quadrature oscillators using CMOS active transformers are also studied.

Chapter 7 presents the theory of current-mode phase-locked loops (PLLs) and examines the intrinsic differences between voltage-mode and current-mode PLLs. The chapter starts with an in-depth study of the configurations and characteristics of voltage-mode PLLs. Both type I and type II voltage-mode PLLs are studied. It then moves on to investigate current-mode PLLs with CMOS active inductors and transformers. The loop dynamics of these PLLs are investigated in detailed. Three design examples are utilized to demonstrate the performance of current-mode PLLs with active inductors and active transformers.

The materials of the book are presented with an emphasis on both the evolution of each class of circuits and a close comparison of their advantages and limitations. The examples given in the book were implemented in TSMC-

0.18μm 1.8V and UMC-0.13μm 1.2V CMOS technologies, and analyzed using SpectreRF from Cadence Design Systems with BSIM3v3RF device models that account for both the parasitics and high-order effects of MOS devices at high frequencies. Readers are assumed to be familiar with the fundamentals of electrical networks, microelectronic devices and circuits, signals and systems, and basic RF circuits.

This book is the first text that provides a comprehensive treatment of the principle, implementation, and applications of CMOS active inductors and transformers. It is a valuable resource for senior undergraduate / graduate students and an important reference for IC design engineers.

Although an immense amount of effort has been made in preparation of the manuscript, flaws and errors will still exist due to erring human nature. Suggestions and corrections will be gratefully appreciated by the author.

FEI YUAN

DECEMBER 31, 2007

Acknowledgments

I would like to take this opportunity to express my sincere gratitude to the Natural Science and Engineering Research Council of Canada, Ryerson University, and CMC Microsystems Inc., Kingston, Ontario, Canada, for their financial and technical supports to our research on integrated circuits and systems. The support from the Department of Electrical and Computer Engineering of Ryerson University where I introduced and taught graduate courses EE8501 (CMOS analog integrated circuits), EE8502 (VLSI systems), and EE8503 (VLSI circuits for data communications) is gratefully acknowledged. I am also grateful to Ryerson University for awarding me the Ryerson Research Chair with both a much needed research grant and a reduced teaching load in 2005-2007 during which much of the research work on active inductors and transformers was carried out. The sabbatical leave from September 2007 to August 2008 provided me with the critically needed time and freedom to complete the manuscript of the monograph.

Special thanks go to my current graduate students Adrian Tang and Dominic DiClemente, and my former graduate students Jean Jiang (Intel Corp., Folsom, CA.), Alec Li (Micron Technologies, Bois, Idaho), and Tao Wang (McMaster University, Hamilton, Canada) for fruitful and productive discussion in our weekly research meetings where many of the original ideas on CMOS active inductors, CMOS active transformers, and their applications in wireless communications and high-speed data communications emerged. Mr. Jason Naughton, our System Administrator, deserves a special thank-you for his prompt response to our random calls on computer/CAD-tool related issues and for keeping CAD tools up-to-date and running all the time.

The editorial staff of Springer, especially Mr. Alex Greene, the Editorial Director of Engineering, have been warmly supportive from the submission of the initial proposal of the book to the completion of the manuscript. Ms. Jennifer Mirski, the Editorial Assistant of Engineering at Springer, deserves a special thank-you for her warm and professional assistance in arranging the review of the submitted manuscript, the design of the lovely cover of the book, and the coordination of the publishing of the book.

Finally and most importantly, this book could not have been possible without the support of my family. I am indebted to my wife Jing for her love, patient, and understanding throughout the project. I also want to thank our daughter and son, Michelle and Jonathan, for the joy that they have brought to our life, and for their forbearance of my bad temper due to the stress of writing.

Contents

PART I

PRINCIPLE AND IMPLEMENTATION OF CMOS ACTIVE INDUCTORS & TRANSFORMERS

Chapter 1

INTRODUCTION

CMOS spiral inductors and transformers have found a broad range of applications in high-speed analog signal processing including impedance matching and gain-boosting in wireless transceivers, bandwidth improvement in broadband data communications over wire and optical channels, oscillators and modulators, RF bandpass filters, RF phase shifters, RF power dividers, and coupling of high-frequency signals, to name a few. Traditionally, passive inductors and transformers are off-chip discrete components. The need for off-chip communications with these passive components severely limits the bandwidth, reduces the reliability, and increases the cost of systems. Since early 1990s, a significant effort has been made to fabricate inductors and transformers on a silicon substrate such that an entire wireless transceiver can be integrated on a single substrate monolithically. In the mean time, the need for a large silicon area to fabricate spiral inductors and transformers has also sparked a great interest in and an intensive research on the synthesis of inductors and transformers using active devices, aiming at minimizing the silicon consumption subsequently the fabrication cost and improving the performance.

This chapter looks into the characteristics of spiral and active inductors and transformers, both their advantages and limitations, and the impact of these characteristics on the applications of these devices. Section 1.1 demonstrates the critical need for an inductive characteristic in high-speed applications. The characteristics of spiral inductors and transformers are examined in Section 1.2. In Section 1.3, we investigate the pros and cons of active inductors and transformers. The chapter is summarized in Section 1.4.

1.1 Inductive Characteristics in High-Speed Applications

Inductive characteristics are critically needed in high-speed applications to improve the performance of the systems, such as improving bandwidth and boosting gain, and to perform specific tasks, such as impedance matching and frequency selection. These applications include LC tank oscillators, bandwidth enhancement in broadband communications, impedance matching in narrow-band communications, phase shifting for RF antennas and radars, RF power dividers, frequency selection, in particular, RF bandpass filters, RF power amplifiers, and gain boosting of RF low-noise amplifiers.

1.1.1   LC Oscillators

One of the key applications of inductors and transformers in wireless communications is the construction of LC oscillators. As compared with ring oscillators, LC oscillators with spiral inductors or transformers offer the key advantage of a low level of phase noise. They are widely used in wireless communication systems where a stringent constraint on the phase noise of oscillators exists. Frequency tuning of LC oscillators with spiral inductors or transformers is typically done by varying the capacitance of the LC tanks as the inductance tuning of spiral inductors or transformers in a monolithic integration is rather difficult. These variable capacitors are usually realized using MOS varactors and offer a relatively small capacitance tuning range, subsequently a small frequency tuning range of the oscillators. Because the need for the frequency tuning range in narrow-band wireless applications is much relaxed as compared with that for clock and data recovery in broadband data communications over wire lines or optical channels, the capacitance-based frequency tuning usually provides an adequate frequency tuning range. Table 1.1 compiles some of the recently published work on LC oscillators with either spiral inductors or spiral transformers.

1.1.2   Bandwidth Improvement

Bandwidth is of a critical concern in broadband communications, such as optical front-ends and data communications over wire links. The bandwidth of a circuit is set by the time constant of the critical node, i.e. the node that has the largest time constant, of the circuit. Three approaches, namely *inductive peaking*, *current-mode signaling*, and *distributed amplification*, are widely used to improve the bandwidth of circuits.

- *Inductive peaking* - The idea is to place an inductor at the node where a large nodal capacitance exists such that the first-order RC network associated with the node is replaced with a second-order RLC network. Because a RLC network has three different modes of operation, namely over damped, critically damped, and under damped. The bandwidth in these three cases differs with under-damped RLC systems exhibit the largest bandwidth. Both shunt peaking [11, 12] and series peaking [13], have been used, as shown

Table 1.1.   Performance comparison of CMOS spiral inductor / transformer LC VCOs.

Reference	Tech.	f_o [GHz]	Tuning range	Phase [dBc/Hz]	Power [mW]
Straayer *et al.* [1] (2002)	0.35μm	1.7	6.3%	-138@0.6MHz	35*
Baek *et al.*[2] (2003)	0.18μm	8.08-7.83	3%	-108@1MHz	24*
Wang *et al.* [3] (2003)	0.18μm	5.75-5.95	3.4%	-105@5MHz	18*
Gierkink *et al.*[4] (2003)	0.25μm	4.57-5.21	13%	-124@1MHz	21.9*
Choi *et al.*[5] (2004)	0.25μm	1.5	–	-100@1MHz	28.8*
Dehghani & Atarodi [6] (2004)	0.25μm	5.8	4.4%	-134.6@3MHz	5.8
Kim *et al.*[7] (2004)	0.25μm	2.27	–	-137@1MHz	30*
Chang & Kim [8] (2005)	0.18μm	5.5-6.7	20%	-115@1MHz	5.76
Oh & Lee [9] (2005)	0.18μm	11	2.73%	-109.4@1MHz	6.84
Soltanian & Kinget [10] (2006)	0.25μm	1.755-2.123	19%	-120@0.6MHz	2.25

1. Legends : TVCO - Transformer VCO; QVCO - Quadrature VCO.
2. VCO proposed by Choi *et al.* is a source-injection quadrature VCO.
3. VCO proposed by Chang and Kim is a cascode-coupling quadrature VCO.
4. VCO proposed by Gierkink *et al.* is a super-harmonic coupled VCO.
5. VCO proposed by Oh and Lee is a back-gate transformer feedback VCO.
6. VCO proposed by Soltanian and Kinget is a tail-current shaped VCO.
7. * Total power consumption including buffers. Otherwise, the power consumption of oscillator core only.

in Fig.1.1. It was demonstrated in [11, 12] that inductive shunt peaking can improve the bandwidth of a common-source amplifier by as much as 70%.

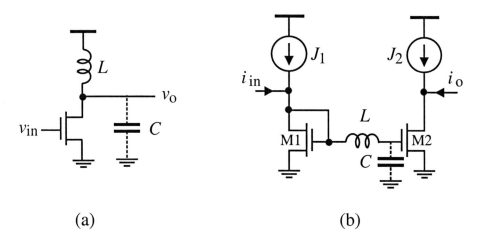

(a)   (b)

Figure 1.1.   Inductive peaking. (a) Shunt peaking - The peaking inductor L is in parallel with the dominant capacitor C. (b) Series peaking - The peaking inductor L is in series with the dominant capacitor C.

- *Current-mode signaling* - A circuit is classified as current-mode if the information carriers of the circuit are the branch currents of the circuit. A key distinction between current-mode and voltage-mode circuits is that the nodal impedance of current-mode circuits is low whereas that of voltage-mode circuits is high. This distinct characteristic forms the foundation on which current-mode circuits are designed. Since the information processed by a current-mode circuit is represented by the branch currents rather than nodal voltages of the circuit, the swing of the nodal voltages of the circuit can be made small without sacrificing the fidelity of the signal. Current-mode circuits offer an improved bandwidth due to the following reasons : (i) Low nodal impedances - the low nodal impedances of current-mode circuits lower the nodal time constants of the circuits. (ii) Low voltage swing - the small swing of the nodal voltages of current-mode circuits reduces the amount of the time required to charge and discharge the nodes of the circuits [14]. It should be emphasized that the speed improvement using current-mode signaling is often moderate as lowering the nodal impedance of a nodal is often echoed with an increase in the capacitance of the node at the same time. Because each node is essentially a first-order RC network whose time constant is given by $\tau_n = R_n C_n$, where R_n and C_n denote the resistance and capacitance of the node, respectively, the net reduction in the nodal time constant is rather moderate.

- *Distributed amplification* - As pointed out earlier that the bandwidth or speed of a circuit is set by the time constant of the critical node of the circuit. An effective way to minimize the effect of the large shunt capacitance of the critical node is to break the large shunt capacitor into several smaller shunt capacitors and separate them with inductors such that the large shunt capacitor is replaced with a distributed LC network or a transmission line [15, 16]. Shown in Fig.1.2 is a common-source amplifier where an shunt-peaking inductor is employed at the drain of the transistor to offset the effect of the large output capacitance C arising from the large width of the transistor. Inductive shunt-peaking, though effective, can not deliver the needed bandwidth in this case. Notice that a transistor is typically laid out in a multi-finger fashion. This is equivalent to connecting N smaller transistors whose width is only $1/N$ of that of the original transistor in parallel, as shown in the figure, where N is the number of the fingers of the transistor. Inductors can then be employed to separate these small transistors and form two transmission lines, one at the drain and the other at the source. Resistors R_{1-4} are for the purpose of impedance matching.

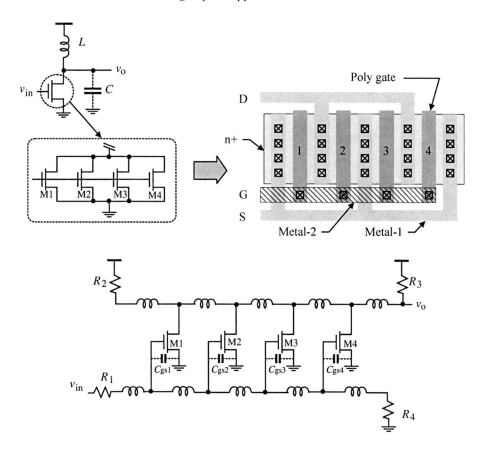

Figure 1.2. Distributed amplifiers.

1.1.3   Impedance Matching

Impedance matching is required between channels and high-speed circuits to minimize signal reflection at their interfaces. Resistors are usually used to provide a matching impedance in broadband communication systems as impedance matching is required over a broad frequency spectrum, as shown in Fig.1.3 where a shunt resistor termination at the far end of the channels is employed. Active termination where transistors are used as termination devices has also been used recently to take the advantage of the tunability of the termination resistance [14]. Frequency-dependent elements, such as capacitors and inductors, can not be used for impedance matching in broadband communication systems simply due to their frequency-dependent characteristics.

Most wireless communication systems operate in a narrow-band mode. A matching impedance in this case is only required over a very narrow frequency

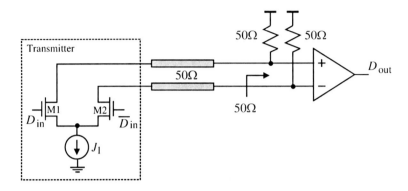

Figure 1.3. Impedance matching in high-speed current-mode serial links.

range. Although resistors can be used for these applications, the high level of the thermal noise of resistors disqualifies themselves from being used in wireless communications where a stringent constraint on the noise performance of the systems exists. Instead, noiseless elements such as capacitors and spiral inductors are widely used in narrow-band impedance matching because these frequency-dependent noiseless elements can provide the desired impedance in a narrow frequency band and at the same time keep the noise at the minimum [17, 18]. Shown in Fig.1.4 is a widely used termination scheme for narrow-band low-noise amplifiers. Neglecting C_{gd} and other parasitic capacitances, it is elementary to show that the input impedance of the LNA is given by

$$z_{in} = \left[j\omega(L_1 + L_2) + \frac{1}{j\omega C_{gs1}} \right] + \frac{g_{m1}L_2}{C_{gs1}}, \qquad (1.1)$$

where C_{gs1} and g_{m1} are the gate-source capacitance and transconductance of M_1, respectively. It is seen from (1.1) that the first term on the right hand side of (1.1) is reactive while the second term is resistive. The reactive term can be made zero by imposing

$$\omega(L_1 + L_2) - \frac{1}{\omega C_{gs1}} = 0. \qquad (1.2)$$

The input impedance of the LNA in this case becomes purely resistive and is given by

$$z_{in} = \frac{g_{m1}L_2}{C_{gs1}}. \qquad (1.3)$$

The need for two inductors L_1 and L_2 with L_1 at the gate and L_2 at the source is justified as the followings : Once the dimension of M_1 is chosen, g_{m1} and

C_{gs1} are determined. The desired input impedance of the LNA in this case can be obtained by adjusting L_2. Once L_2 is chosen, the value of L_1 can be tuned to ensure the total cancellation of the reactive part of the input impedance.

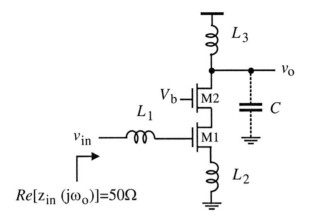

Figure 1.4. Impedance matching in narrow-band low-noise amplifiers.

1.1.4 Phase Shifting

A phase shifter is a uni-directional serial network inserted in a signal path so that the phase of the signal at the output of the signal path can be adjusted in a controlled manner [19, 20]. A well-designed phase shifter should possess the characteristics of a low insertion loss, a high return loss, and a large phase shift range. The common configuration of RF phase shifters is shown in Fig.1.5. The floating inductors are typically implemented using high-impedance metal lines. The tuning of the amount of the phase shift is carried out by varying the capacitance of the shunt varactors.

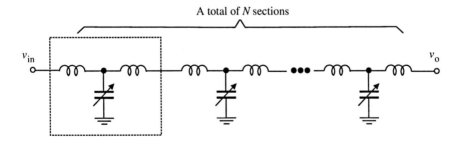

Figure 1.5. RF phase shifters with floating inductors and shunt varactors.

1.1.5   Frequency Selection

Bandpass filters with a high passband center frequency are used extensively in narrow-band wireless communications for RF band selection. These filters are traditionally implemented using lumped LC and surface acoustic wave (SAW) filters with SAW filters the most widely used. SAW filters, however, are not compatible with CMOS technologies. The recent effort on integrating RF bandpass filters on a silicon substrate is accelerated with the emergence of CMOS spiral inductors and transformers. A single-chip realization of RF transceivers with on-chip RF bandpass filters offers a number of critical advantages including a reduced assembly cost, increased system reliability, and improved performance. To compensate for the high insertion loss of spiral inductors at high frequencies, active negative resistors have been used to boost the quality factor of the spiral inductors. Table 1.2 tabulates some recently reported RF bandpass filters with spiral inductors and active Q-enhancement.

Table 1.2.   Bandpass filter with Q-enhanced spiral inductors.

Reference	Year	Tech.	f_o	Tuning range
Kuhn *et al.* [21]	1997	2μm	0.2 GHz	4.5%
Duncan *et al.* [22]	1997	0.8μm	0.75 GHz	2.3%
Soorapanth-Wong [23]	2002	0.25μm	2.14 GHz	1.4%
Dulger *et al.* [24]	2003	0.35μm	2.1 GHz	13%
Kuhn *et al.* [25]	2003	-	0.9 GHz	2.2%
Bantas-Koutsoyannopoulos [26]	2004	0.35μm	1 GHz	11%
Kulyk-Haslett [27]	2006	0.18μm	2.368 GHz	1.3%

1.1.6   Gain Boosting

Traditional gain-boosting techniques such as cascodes and regulated cascodes lose their potency at high frequencies due to the increased gate-source and gate-drain couplings via the gate-source and gate-drain capacitors of MOSFETs. A technique that is widely used in boosting the voltage gain of narrowband low-noise amplifiers (LNAs) is to use a LC tank as the load of the LNAs, utilizing the infinite impedance of ideal LC tanks at their self-resonant frequency. When a LC tank is used as the load of a common-source amplifier whose voltage gain is approximated by $A_v \approx -g_m z_L$, where g_m is the transconductance of the MOSFET and z_L is the load impedance, as shown in Fig.1.6, the large impedance of the LC tank at its self-resonant frequency $\omega_o = \dfrac{1}{\sqrt{L_p C}}$

will significantly boost the gain of the amplifier at ω_b. The resonant frequency of the tank is set to be the same as the frequency of the input of the LNA. Note that voltage gain of the amplifier at frequencies other than ω_b is low.

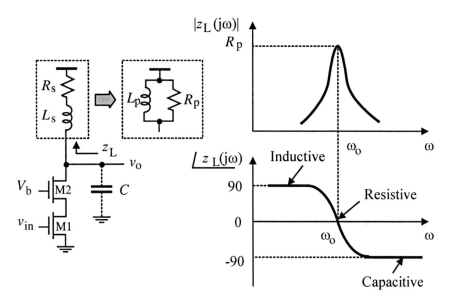

Figure 1.6. Gain boosting of low noise amplifiers using LC tank load.

1.1.7 Power Dividers

RF power dividers are traditionally realized using transmission lines. Lumped passive implementation of power dividers has also been used to reduce the size of the power dividers, however, at the cost of a high insertion loss and a limited bandwidth. Fig.1.7 shows the equivalent circuit of lumped Wilkinson power divider. The use of CMOS active inductors to replace passive spiral inductors of RF power dividers was proposed by Lu and Wu in [28] to take the advantages of the high quality factor, low silicon consumption, and high self-resonant frequency of CMOS active inductors.

1.2 Spiral Inductors and Transformers

Monolithic on-chip inductors and transformers are also known as spiral inductors and transformers due to the way in which these inductors and transformers are laid out. Both planar and stacked spiral inductors and transformers have been developed and the detailed characterization and modeling of these inductors and transformers are available. Modern CAD tools for IC design are equipped with spiral inductors as standard elements in their component libraries.

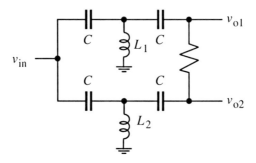

Figure 1.7.   Equivalent circuit of lumped Wilkinson power divider.

1.2.1   Planar Spiral Inductors

Fig.1.8 is a sketch of square-shaped planar spiral inductors. The lumped equivalent circuit of spiral inductors is given in Fig.1.9, where L is the inductance of the spiral inductor, R_s represents the series resistance of the spiral caused by the skin-effect induced resistance in the spiral and the resistance induced by the eddy current in the substrate [29], C_s accounts for the capacitance due to the overlap of the spiral and the center-tap underpasses, C_{ox} denotes the capacitance between the spiral and the substrate, C_b and R_b quantify the capacitance and resistance of the substrate, respectively. Although modern CMOS technologies are equipped with multiple metal layers, typically only the top metal layer is used to construct planar spiral inductors and transformers such that the unwanted parasitic capacitance between the spiral and the substrate is minimized. As pointed out in [29], the substrate loss accounts for 10-30% reduction of the quality factor of the spiral inductors in low GHz ranges, mainly due to the penetration of the electric field generated by the spiral into the substrate. A main drawback of planar spiral inductors is their low inductance.

1.2.2   Stacked Spiral Inductors

The inductance of spiral inductors can be increased significantly using stacked configurations, as shown in Fig.1.10, however, at the price of an increased spiral-substrate capacitance because the lower metal layers are also used in the construction of the inductors [30]. The total inductance of a stacked inductor with two spiral layers is given by

$$L_{total} = L_1 + L_2 + 2M, \tag{1.4}$$

where L_1 and L_2 are the self-inductances of spirals 1 and 2, respectively, and M is the mutual inductance between the two spirals. Note the direction of the routing of the spirals in differential metal layers must be carefully

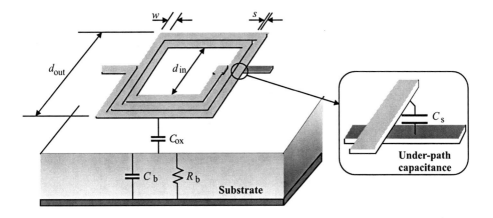

Figure 1.8. Square-shaped planar spiral inductors. w is the width of the spiral and s is the spacing between the turns of the spiral.

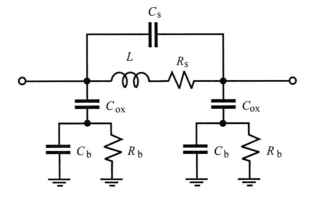

Figure 1.9. Lumped circuit model of planar spiral inductors.

chosen such that the total inductance increases. Figs.1.11 and 1.12 show the dependence of the measured inductance and self-resonant frequency of stacked inductors on the number of spiral layers [30]. It is seen that the inductance of stacked spiral inductors increases approximately linearly with the increase in the number of the spiral layers of the inductors. The self-resonant frequency of stacked inductors decreases with the increase in the number of spiral layers in a nonlinear fashion.

1.2.3 Spiral Transformers

To increase the inductance without significantly increasing silicon consumption, stacked spiral transformers and bilifar spiral transformers emerged. Transformers are essentially two spiral inductors coupled via a magnetic link.

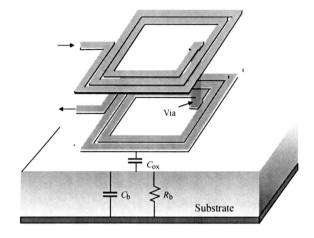

Figure 1.10.   Square-shaped stacked spiral inductors.

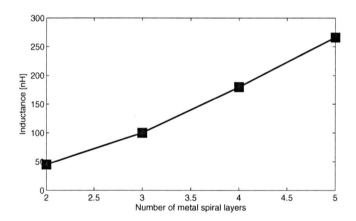

Figure 1.11.   Dependence of the inductance of stacked spiral inductors on the number of spiral layers.

Fig.1.13 shows the lumped circuit model of bifilar spiral transformers [31], where C_{12} is the mutual capacitance between the primary and secondary windings, and k_{12}, k_{21} are the coupling coefficients. All other parameters are the same as those in Fig.1.9.

1.2.4   Characteristics of Spiral Inductors and Transformers

Spiral inductors and transformers offer the key advantages of superior linearity and a low level of noise. The performance and applications of spiral inductors and transformers are affected by a number of drawbacks that are intrinsic to the physical geometry of these passive devices and CMOS tech-

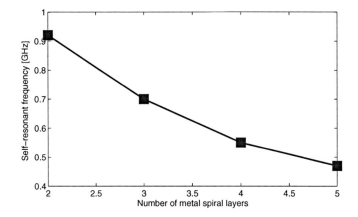

Figure 1.12. Dependence of the self-resonant frequency of stacked spiral inductors on the number of spiral layers.

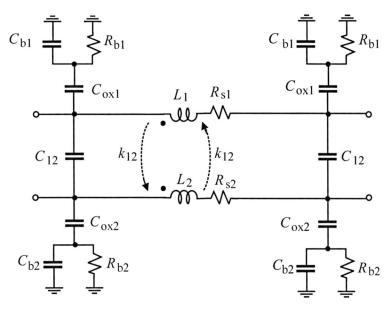

Figure 1.13. Lumped circuit model of spiral transformer.

nologies in which spiral inductors and transformers are implemented. These drawbacks include :

- *Low quality factor* - The quality factor of spiral inductors and transformers is limited by the ohmic loss of the spiral at high frequencies. Two sources that contribute to the ohmic loss of the spiral inductors and transformers exist : the skin-effect induced resistance of the spiral and the resistance induced by the eddy currents in the substrate. The former is proportional to the square-root of the frequency of the signal flowing in the spiral whereas the latter is proportional to the square of the frequency of the signal flowing in the spiral. Because the resistance of the substrate in the lateral direction is much larger as compared with that in the vertical direction [32], the eddy currents in the literal direction in the substrate are small. The substrate eddy-current induced resistance of the spiral is often negligible as compared with the resistance caused by the skin-effect [23]. The quality factor of spiral inductors in standard CMOS is typically below 20, as is evident in Table 1.3 where the parameters of some recently reported CMOS spiral inductors are tabulated.

- *Low self-resonant frequency* - The self-resonance of a spiral inductor is the resonance of the LC tank formed by the series inductance of the spiral inductor and the shunt capacitance between the spiral of the inductor and the substrate, as well as its underpass capacitance. The low self-resonant frequency of spiral inductors is mainly due to the large spiral-substrate capacitance, arising from the large metal area occupied by the spiral. The typical self-resonant frequency of spiral inductors is in the low GHz range, as is evident in Table 1.3.

- *Low and fixed inductance* - As pointed out earlier that the inductance of a planar spiral inductor is low and fixed once the number of the turns of the spiral is set. The only way to increase the inductance of the spiral inductor is to either increase the number of the turns of the spiral or use a stacked configuration where spirals on multiple metal layers are connected using vias. The former is at the expense of a large silicon area while the latter increases the spiral-substrate capacitance.

- *Large silicon area* - Due to the low inductance of spiral inductors, especially planar spiral inductors, and the fact that the inductance of these inductors is directly proportional to the number of the turns of the spiral of the inductors, the silicon area required for routing the spiral of the inductors is large. As an example, it was shown in [11] that a square-shaped spiral inductor of an inductance 8 nH will require 6 turns with $d_{out} = 400\mu m$. The actual silicon consumption of the inductor is larger than $400 \times 400 \mu m^2$ when design rules on the minimum spacings are followed. Moreover, the design rules of most IC foundries prohibit the placement of either active or passive devices in the region between the spiral of the inductors and the substrate, making spiral inductors the most silicon-consuming components.

Table 1.3. Key parameters of some CMOS spiral inductors.

Structure	Inductance [nH]	d_{out} [μm]	Turns	$Q(\omega_o)$	f_o [GHz]	Year	Ref.
Square	3.5	330	4	11	3.9	1996	[21]
Square	<6	–	7	<6	1	1996	[33]
Square	15.2	310	8	<4	1.5	1998	[34]
Square	8	400	6	–	–	1999	[11]
Octagon	6	320	4	11	1.4	2003	[35]
Octagon	4.3	200	3	5	4	2003	[36]

1.3   Active Inductors and Transformers

CMOS active inductors are active networks that consist mainly of MOS transistors. Resistors are sometimes used as feedback elements to improve the performance of active inductors. Under certain dc biasing conditions and signal-swing constraints, these networks exhibit an inductive characteristic in a specific frequency range. As compared with their spiral counterparts, CMOS active inductors offer the following attractive advantages :

- *Low silicon consumption* - Because only MOS transistors are usually required in the realization of CMOS active inductors and the inductance of these active inductors is inversely proportional to the transconductances of the transistors, the silicon consumption of CMOS active inductors is negligible as compared with that of their spiral counterparts.

- *Large and tunable self-resonant frequency* - CMOS active inductors with a large self-resonant frequency are highly desirable. For example, the passband center frequency of an active inductor RF bandpass filter is typically set to the self-resonant frequency of the active inductor of the filter. The larger the self-resonant frequency of the active inductor, the higher the passband center frequency of the filter. In applications where CMOS active inductors are used for low-pass filters, these filters are operated at frequencies below their self-resonant frequency of active inductors. A large self-resonant frequency of active inductors ensures that the active inductors will have an inductive characteristic over a large frequency range. As to be seen in Chapter 2 that the self-resonant frequency of a CMOS active inductor is the maximum frequency of the transconductors constituting the active inductor. When the basic configurations of transconductors, such as common-source and common-gate transconductors, are used, this frequency approaches f_T of the devices.

- *Large and tunable inductance* - As to be seen in Chapter 2, the inductance of CMOS active inductors is inversely proportional to the transconductances of the transistors synthesizing the inductors. The smaller the width of the transistors, the larger the inductance. Also the inductance can be tuned conveniently by varying the dc biasing condition of the transistors synthesizing the inductor with a large inductance tuning range. The coarse tuning of the inductance of active inductors is typically attained in this way. The fine tuning of the inductance of active inductors can also be achieved by varying the load capacitance of the transconductors of the active inductors using MOS varactors.

- *Large and tunable quality factor* - The quality factor of CMOS active inductors is set by the ohmic loss of the inductors, arising mainly from the finite output resistance of the transconductors of the inductors. The quality factor of CMOS active inductors can be increased by increasing this output resistance. A number of methods are available to boost the output resistance, such as cascodes, regulated cascodes, and negative resistor compensation. In each of these approaches, the degree of compensation can be varied. For example, in the cascode approach, the output resistance of a cascode-configured transconductor can be adjusted by varying the biasing voltage of the cascoding transistor. In the regulated cascode approach, the output resistance of the regulated-cascode configured transconductor can be changed by varying the voltage gain of its auxiliary voltage amplifier. In the negative resistor compensation approach, the resistance of the compensating negative resistor can be adjusted by varying the biasing current of the negative resistor. A detailed presentation of these approaches will be given in Chapter 2.

- *Compatibility with digital CMOS technologies* - Spiral inductors are not available in low-cost digital-oriented CMOS processes. They are available only in more expensive mixed-mode CMOS technologies. CMOS active inductors, however, can be realized using standard digital CMOS processes.

CMOS active inductors and transformers have found increasing applications in areas where an inductive characteristic is required. These applications include LC and ring oscillators, RF bandpass filters, RF phase shifters, limiting amplifiers for optical communications, low-noise amplifiers for wireless communications, RF power dividers, ultra wideband low-noise amplifiers, and transceivers for high-speed data links over wire lines. Table 1.4 summarizes some of the recently published work where CMOS active inductors and transformers were employed.

Table 1.4. Applications of CMOS active inductors and transformers.

Reference	Year	Tech.	Applications	Remarks
Thanachayanont & Payne [37]	2000	0.8μm	IF Bandpass	100 MHz
Thanachayanont [38]	2000	0.8μm	LC VCO	0.45-1.2 GHz
Lin & Payne [39]	2000	0.35μm	LC VCO	1.1-2.1 GHz
Säckinger & Fischer [40, 41]	2000	0.25μm	Limiting amp.	3 GHz
Wu *et al.* [42]	2001	0.35μm	LC VCO	0.1-0.9 GHz
Grözing *et al.*[43, 44]	2001	0.30μm	LC VCO	0.4-4 GHz
Wu *et al.* [45, 46, 36]	2001	0.35μm	RF bandpass	0.4-1.1 GHz
Thanachayanont [47]	2002	0.35μm	RF bandpass	2.4-2.6 GHz
Xiao & Schaumann [48]	2002	0.18μm	RF lowpass	4.57 GHz
Xiao *et al.* [49]	2004	0.18μm	RF bandpass	3.5-5.7 GHz
Lu *et al.* [28]	2005	0.18μm	RF power divider	4.5 GHz
Liang *et al.* [50]	2005	0.18μm	RF bandpass	3.45-3.6 GHz
Gao *et al.* [51–53]	2005	0.25μm	RF bandpass	2.05-2.45 GHz
Chen *et al.* [54, 55]	2005	0.35μm	Limiting amp.	2.3 GHz
Lu & Liao [20]	2005	0.18μm	RF Phase shifter	360 degrees
Mahmoudi & Salama [56]	2005	0.18μm	QVCO	8 GHz
Jiang & Yuan [57]	2005	0.18μm	Gb/s Tx.	10 Gbps
Yuan [58, 59]	2006	0.18μm	Ring VCO	1.7-2.7 GHz
Lu *et al.* [60]	2006	0.18μm	LC VCO	0.5-3.0 GHz
Xiao & Schaumann [61]	2007	0.18μm	RF bandpass	3.34-5.72 GHz
Weng & Kuo [62]	2007	0.18μm	RF bandpass	2-2.9 GHz
Tang *et al.* [63]	2007	0.18μm	RF modulators	1.6 GHz
Tang *et al.* [64]	2007	0.18μm	VCOs	1.6 GHz

1. For bandpass filters, the frequency range is the passband center frequency range.
2. For VCOs, the frequency range is the oscillation frequency range.
3. For limiting amplifiers, the frequency is the bandwidth of the amplifiers.

The applications of active inductors, however, are affected by several difficulties arising from the intrinsic characteristics of MOS devices. These difficulties include a limited dynamic range, a high level of noise, and a high sensitivity to process spread, supply voltage fluctuation, and ground bouncing. It should be noted that these limitations are not unique to CMOS active inductors but rather common to all synthesized devices. Also, the effect of many of these difficulties can be reduced through innovative designs. For example, the limited dynamic range of active inductors can be expanded using class AB configurations where the voltage swing of active inductors can be made nearly rail-to-rail [65]. The effect of the high sensitivity to process spread can be minimized by making use of the tunability of both the inductance and quality factor of active inductors. The effect of supply voltage fluctuation and ground bouncing can be greatly reduced by using replica-biasing techniques and proper circuit configurations

[66]. The effect of the high level of the noise of active inductors on the phase noise of LC oscillators with active inductors can be minimized by boosting the quality factor of the active inductors [67].

New design techniques are yet critically needed to further improve the performance of active inductors and transformers. The emerging applications of CMOS active inductors continue to be unfolded along with the inception of new design techniques and circuit topologies of these active devices.

1.4   Chapter Summary

The critical need for inductive characteristics in high-speed applications, specifically, LC oscillators, bandwidth improvement of broadband amplifiers, narrow-band impedance matching, RF phase shifters, RF power dividers, RF frequency selection, and gain boosting of LNAs, has been investigated. An in-depth evaluation of the pros and cons of spiral inductors and transformers has been given. We have shown that spiral inductors and transformers suffer from a low quality factor, a low self-resonant frequency, a low and fixed inductance, and the need for a large silicon area.

We have shown that CMOS active inductors that are synthesized using MOS transistors offer a number of attractive characteristics as compared with their spiral counterparts. These characteristics include a low silicon consumption, a large and tunable self-resonant frequency, a large and tunable inductance, a large and tunable quality factor, and fully realizable in digital CMOS technologies. The applications of active inductors, however, are affected by several stiff difficulties arising from the intrinsic characteristics of MOS devices including a limited dynamic range, a high level of noise, and a high sensitivity to process spread, supply voltage fluctuation, and ground bouncing.

Chapter 2

CMOS ACTIVE INDUCTORS

This chapter provides an in-depth treatment of the principles, topologies, characteristics, and implementation of CMOS active inductors. Section 2.1 investigates the principles of gyrator-C based synthesis of inductors. Both single-ended and floating (differential) configurations of gyrator-C active inductors are studied. Section 2.2 investigates the most important figure-of-merits that quantify the performance of active inductors. These figure-of-merits include frequency range, inductance tunability, quality factor, noise, linearity, stability, supply voltage sensitivity, parameter sensitivity, signal sensitivity, and power consumption. Section 2.3 details the CMOS implementation of single-ended gyrator-C active inductors. The schematics and characteristics of floating gyrator-C active inductors are examined in detail in Section 2.4. Class AB active inductors are investigated in Section 2.5. The chapter is summarized in Section 2.6.

2.1 Principles of Gyrator-C Active Inductors

2.1.1 Lossless Single-Ended Gyrator-C Active Inductors

A gyrator consists of two back-to-back connected transconductors. When one port of the gyrator is connected to a capacitor, as shown in Fig.2.1, the network is called the gyrator-C network. A gyrator-C network is said to be lossless when both the input and output impedances of the transconductors of the network are infinite and the transconductances of the transconductors are constant.

Consider the lossless gyrator-C network shown in Fig.2.1(a). The admittance looking into port 2 of the gyrator-C network is given by

$$Y = \frac{I_{in}}{V_2} = \frac{1}{s\left(\dfrac{C}{G_{m1}G_{m2}}\right)}. \tag{2.1}$$

Eq.(2.1) indicates that port 2 of the gyrator-C network behaves as a single-ended lossless inductor with its inductance given by

$$L = \frac{C}{G_{m1}G_{m2}}. \tag{2.2}$$

Gyrator-C networks can therefore be used to synthesize inductors. These synthesized inductors are called gyrator-C active inductors. The inductance of gyrator-C active inductor is directly proportional to the load capacitance C and inversely proportional to the product of the transconductances of the transconductors of the gyrator. Also, the gyrator-C network is inductive over the entire frequency spectrum. It should also be noted that the transconductor in the forward path can be configured with a negative transconductance while the transconductor in the feedback path has a positive transconductance, as shown in Fig.2.1(b).

Although the transconductors of gyrator-C networks can be configured in various ways, the constraint that the synthesized inductors should have a large frequency range, a low level of power consumption, and a small silicon area requires that these transconductors be configured as simple as possible. Fig.2.2 shows the simplified schematics of the basic transconductors that are widely used in the configuration of gyrator-C active inductors. Common-gate, common-drain, and differential-pair transconductors all have a positive transconductance while the common-source transconductor has a negative transconductance. To demonstrate this, consider the common-gate transconductor. An increase in v_{in} will lead to a decrease in i_D. Because $i_o = J - i_D$, i_o will increase accordingly. So the transconductance of the common-gate transconductor is positive. Similarly, for the differential-pair transconductor in Fig.2.2(d). An increase in v_{in} will result in an increase in i_{D1}. Since $i_{D2} = J_3 - i_{D1}$, i_{D2} will decrease. Further $i_o = J_2 - i_{D2}$, i_o will increase. The differential-pair transconductor thus has a positive transconductance.

2.1.2 Lossless Floating Gyrator-C Active Inductors

An inductor is said to be floating if both the terminals of the inductor are not connected to either the ground or power supply of the circuits containing the active inductor. Floating gyrator-C active inductors can be constructed in a similar way as single-ended gyrator-C active inductors by replacing single-ended transconductors with differentially-configured transconductors, as shown in Fig.2.3. Because

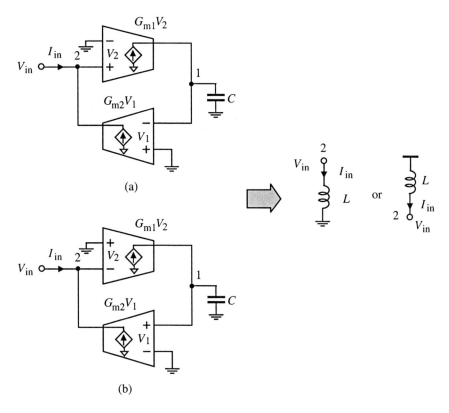

Figure 2.1. Lossless singe-ended gyrator-C active inductors. G_{m1} and G_{m2} are the transconductances of transconductors 1 and 2, respectively, and C is the load capacitance at node 1. (a) Transconductor in the forward path has a positive transconductance while the transconductor in the feedback path has a negative transconductance; (b) Transconductor in the forward path has a negative transconductance while the transconductor in the feedback path has a positive transconductance.

$$V_{in1}^{+} = -\frac{g_{m1}}{sC}(V_{in2}^{+} - V_{in2}^{-}),$$

$$V_{in1}^{-} = \frac{g_{m1}}{sC}(V_{in2}^{+} - V_{in2}^{-}), \qquad (2.3)$$

$$I_{o2} = g_{m2}(V_{in1}^{+} - V_{in1}^{-}),$$

we have

$$I_{o2} = -\frac{2g_{m1}g_{m2}}{sC}(V_{in2}^{+} - V_{in2}^{-}). \qquad (2.4)$$

The admittance looking into port 2 of the gyrator-C network is given by

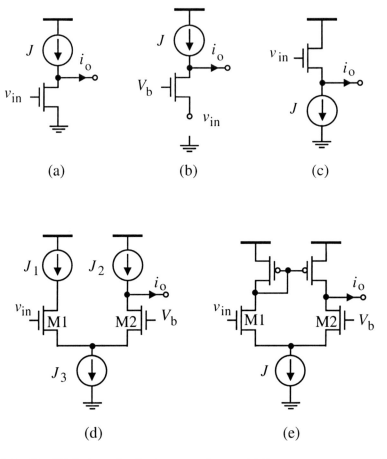

Figure 2.2. Simplified schematic of basic transconductors. (a) Common-source transconductor $i_o = -g_m v_{in}$; (b) Common-gate transconductor $i_o = g_m v_{in}$; (c) Common-drain (source follower) transconductor $i_o = g_m v_{in}$; (d,e) Differential-pair transconductors $i_o = g_m v_{in}$.

$$Y = \frac{I_{in}}{V_{in2}^+ - V_{in2}^-}$$

$$= \frac{1}{s\left(\dfrac{2C}{g_{m1}g_{m2}}\right)}. \tag{2.5}$$

Eq.(2.5) reveals port 2 of the gyrator-C network behaves as a floating inductor with its inductance given by

$$L = \frac{2C}{g_{m1}g_{m2}}. \tag{2.6}$$

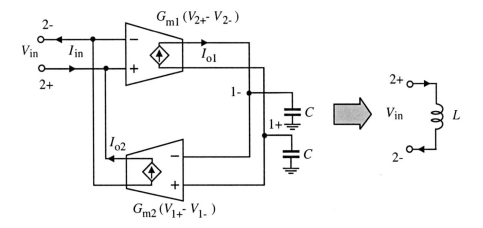

Figure 2.3. Lossless floating gyrator-C active inductors. G_{m1} and G_{m2} are the transconductances of transconductors 1 and 2, respectively, and C is the load capacitance at nodes 1+ and 1-.

Floating gyrator-C active inductors offer the following attractive advantages over their single-ended counterparts : (i) The differential configuration of the transconductors effectively rejects the common-mode disturbances of the network, making them particularly attractive for applications where both analog and digital circuits are fabricated on the same substrate. (ii) The level of the voltage swing of floating active inductors is twice that of the corresponding single-ended active inductors.

2.1.3 Lossy Single-Ended Gyrator-C Active Inductors

When either the input or the output impedances of the transconductors of gyrator-C networks are finite, the synthesized inductors are no longer lossless. Also, the gyrator-C networks are inductive only in a specific frequency range.

Consider the gyrator-C network shown in Fig.2.4 where G_{o1} and G_{o2} denote the total conductances at nodes 1 and 2, respectively. Note G_{o1} is due to the finite output impedance of transconductor 1 and the finite input impedance of transconductor 2. To simplify analysis, we continue to assume that the transconductances of the transconductors are constant. Write KCL at nodes 1 and 2

$$(sC_1 + G_{o1})V_1 - G_{m1}V_2 = 0 \qquad \text{(node 1)},$$

$$-I_{in} + (sC_2 + G_{o2})V_2 - G_{m2}(-V_1) = 0, \quad \text{(node 2)}.$$

(2.7)

The admittance looking into port 2 of the gyrator-C network is obtained from

$$Y = \frac{I_{in}}{V_2}$$

$$= sC_2 + G_{o2} + \frac{1}{s\left(\dfrac{C_1}{G_{m1}G_{m2}}\right) + \dfrac{G_{o1}}{G_{m1}G_{m2}}}. \tag{2.8}$$

Eq.(2.8) can be represented by the RLC networks shown in Fig.2.4 with its parameters given by

$$R_p = \frac{1}{G_{o2}},$$

$$C_p = C_2,$$

$$R_s = \frac{G_{o1}}{G_{m1}G_{m2}}, \tag{2.9}$$

$$L = \frac{C_1}{G_{m1}G_{m2}}.$$

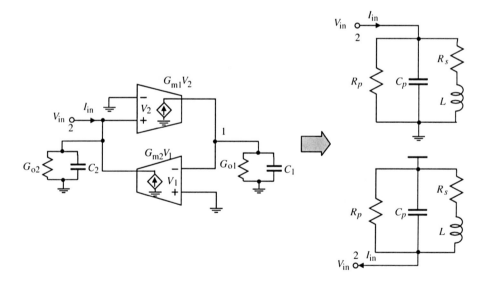

Figure 2.4. lossy single-ended gyrator-C active inductors. C_1 and G_{o1}, C_2 and G_{o2} denote the total capacitances and conductances at nodes 1 and 2, respectively.

We comments on the preceding results :

- When the input and output conductances of the transconductors are considered, the gyrator-C network behaves as a lossy inductor with its parasitic parallel resistance R_p, parallel capacitance C_p, and series resistance R_s. R_p should be maximized while R_s should be minimized to low the ohmic loss. The finite input and output impedances of the transconductors of the gyrator-C network, however, have no effect on the inductance of the active inductor.

- R_p and C_p are solely due to G_{o2} and C_2. G_{o1} and C_1 only affect R_s and L.

- The resonant frequency of the RLC networks of the active inductor is given by

$$\omega_o = \frac{1}{LC_p} = \sqrt{\frac{G_{m1}}{C_1}\frac{G_{m2}}{C_2}} = \sqrt{\omega_{t1}\omega_{t2}}, \tag{2.10}$$

where

$$\omega_{t1,2} = \frac{G_{m1,2}}{C_{1,2}} \tag{2.11}$$

is the cut-off frequency of the transconductors. ω_o is the self-resonant frequency of the gyrator-C active inductor. This self-resonant frequency is typically the maximum frequency at which the active inductor operates. The self-resonant frequency of an active inductor is set by the cut-off frequency of the transconductors constituting the active inductor.

- The small-signal behavior of a gyrator-C active inductor is fully characterized by its RLC equivalent circuit. The RLC equivalent circuit of gyrator-C active inductors, however, can not be used to quantify the large-signal behavior, such as the dependence of the inductance on the dc biasing condition of the transconductors and the maximum signal swing of the gyrator-C active inductors.

- When the conductances encountered at nodes 1 and 2 of the gyrator-C active inductors are zero (lossless), the phase of the impedance of the synthesized inductor is $\frac{\pi}{2}$. However, when these conductances are non-zero, the phase of the impedance of the synthesized inductor will deviate from $\frac{\pi}{2}$, giving rise to a phase error. The phase error is due to R_p and R_s of the active inductors. The phase of the impedance of practical active inductors should be made constant and to be as close as possible to $\frac{\pi}{2}$.

■ The finite input and output impedances of the transconductors constituting active inductors result in a finite quality factor. For applications such as band-pass filters, active inductors with a large quality factor are mandatory. In these cases, Q-enhancement techniques that can offset the detrimental effect of R_p and R_s should be employed to boost the quality factor of the active inductors.

2.1.4   Lossy Floating Gyrator-C Active Inductors

Lossy floating gyrator-C active inductors can be analyzed in a similar way as lossy single-ended gyrator-C active inductors. Consider the lossy floating gyrator-C network shown in Fig.2.5. We continue to assume that the transconductances of the transconductors are constant. Writing KCL at nodes 1-, 1+, 2-, and 2+ yields

$$-G_{m1}(V_2^+ - V_2^-) + \left(\frac{sC_1 + G_{o1}}{2}\right)(V_1^- - V_1^+) = 0,$$

$$I_{in} + \left(\frac{sC_2 + G_{o2}}{2}\right)(V_2^- - V_2^+) + G_{m2}(V_1^+ - V_1^-) = 0,$$

(2.12)

The admittance looking into port 2 of the gyrator-C network is obtained from

$$\begin{aligned}
Y &= \frac{I_{in}}{V_2^+ - V_2^-} \\
&= s\frac{C_2}{2} + \frac{G_{o2}}{2} + \frac{1}{s\left(\dfrac{C_1}{2G_{m1}G_{m2}}\right) + \dfrac{G_{o1}}{2G_{m1}G_{m2}}}.
\end{aligned}$$

(2.13)

Eq.(2.12) can be represented by the RLC network shown in Fig.2.5 with its parameters given by

$$R_p = \frac{2}{G_{o2}},$$

$$C_p = \frac{C_2}{2},$$

$$R_s = \frac{G_{o1}/2}{G_{m1}G_{m2}},$$

$$L = \frac{C_1/2}{G_{m1}G_{m2}}.$$

(2.14)

The constant 2 in (2.14) is due to the floating configuration of the active inductor. The capacitance and conductance at the interface nodes $1+$ and $1-$ and those at the internal nodes $2+$ and $2-$ will become $C_2/2$ and $G_{o2}/2$, and $C_1/2$ and $G_{o21}/2$, respectively. Eq.(2.9) can be used to derive (2.14) without modification.

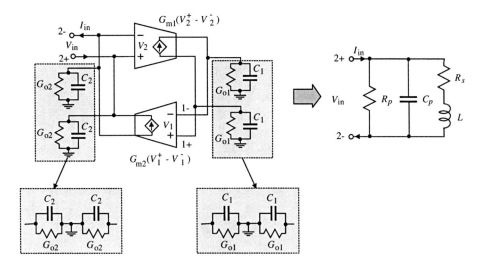

Figure 2.5. Lossy floating gyrator-C active inductors. C_1 and G_{o1}, C_2 and G_{o2} represent the total capacitances and conductances at nodes 1 and 2, respectively.

2.2 Characterization of Active Inductors

In this section, we investigate the most important figure-of-merits that provide quantitative measures of the performance of active inductors. These figure-of-merits include frequency range, inductance tunability, quality factor, noise, linearity, stability, supply voltage sensitivity, parameter sensitivity, signal sensitivity, and power consumption.

2.2.1 Frequency Range

It was shown in the preceding section that an lossless gyrator-C active inductor exhibits an inductive characteristic across the entire frequency spectrum. A lossy gyrator-C active inductor, however, only exhibits an inductive characteristic over a specific frequency range. This frequency range can be obtained by examining the impedance of the RLC equivalent circuit of the lossy active inductor

$$Z = \left(\frac{R_s}{C_pL}\right) \frac{s\frac{L}{R_s} + 1}{s^2 + s\left(\frac{1}{R_pC_p} + \frac{R_s}{L}\right) + \frac{R_p + R_s}{R_pC_pL}}. \tag{2.15}$$

When complex conjugate poles are encountered, the pole resonant frequency of Z is given by

$$\omega_p = \sqrt{\frac{R_p + R_s}{R_pC_pL}}. \tag{2.16}$$

Because $R_p \gg R_s$, Eq.(2.16) is simplified to

$$\omega_p \approx \sqrt{\frac{1}{LC_p}} = \omega_o, \tag{2.17}$$

where ω_o is the self-resonant frequency of the active inductor. Also observe that Z has a zero at frequency

$$\omega_z = \frac{R_s}{L} = \frac{G_{o1}}{C_1}. \tag{2.18}$$

The Bodé plots of Z are sketched in Fig.2.6. It is evident that the gyrator-C network is resistive when $\omega < \omega_z$, inductive when $\omega_z < \omega < \omega_o$, and capacitive when $\omega > \omega_o$. The frequency range in which the gyrator-C network is inductive is lower-bounded by ω_z and upper-bounded by ω_o. Also observed is that R_p has no effect on the frequency range of the active inductor. R_s, however, affects the lower bound of the frequency range over which the gyrator-C network is inductive. The upper bound of the frequency range is set by the self resonant frequency of the active inductor, which is set by the cut-off frequency of the transconductors constituting the active inductor. For a given inductance L, to maximize the frequency range, both R_s and C_p should be minimized.

2.2.2 Inductance Tunability

Many applications, such as filters, voltage or current controlled oscillators, and phase-locked loops, require the inductance of active inductors be tunable with a large inductance tuning range. It is seen from (2.9) that the inductance of gyrator-C active inductors can be tuned by either changing the load capacitance or varying the transconductances of the transconductors constituting the active inductors. Capacitance tuning in standard CMOS technologies is usually done by using varactors. Two types of varactors exists, namely pn-junction varactors

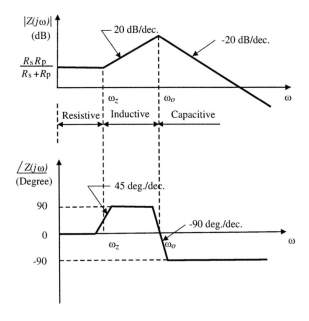

Figure 2.6. Bodé plots of the impedance of lossy gyrator-C active inductors.

and MOS varactors. The sideviews of pn-junction varactors are shown in Fig.2.7 for p+/n-well junctions and Fig.2.8 for n+/p-well junctions. Because p-substrate is connected to the ground, n+/p-well varactors are single-ended. p+/n-well varactors, on the other hand, are floating varactors. The swing of the voltages at the nodes of the varactors must ensure that the n+/p-well and p+/n-well junctions be revise biased all the time such that a junction capacitance exists. The junction capacitance of an abrupt pn-junction is given by

$$C_J = \frac{C_{Jo}}{\sqrt{1 + \dfrac{v_R}{\phi_o}}},$$ (2.19)

where C_{Jo} is the junction capacitance at zero-biasing voltage, v_R is the reverse biasing voltage of the junction and ϕ_o is built-in potential of the junction. It is seen that C_J varies with v_R in a nonlinear fashion. The performance of junction varactors is affected by the following factors :

- Large parasitic series resistance - p+/n-well varactors suffer from a large series resistance - the resistance of the n-well. As a result, the quality factor of the varactor quantified by

$$Q = \frac{1}{\omega R_{n-well} C},$$ (2.20)

where R_{n-well} is the parasitic series resistance, is small. To minimize this unwanted resistance, the spacing between p+ and n+ diffusions should be minimized.

- Large parasitic capacitance between n-well and p-substrate - The larger the capacitance of the varactors, the larger the n-well and subsequently the larger the n-well/p-substrate junction capacitance.

- Small capacitance tuning range - The nonlinear characteristics of C_J result in a small capacitance tuning range with a low capacitance tuning ratio.

- Stringent voltage swing requirement - As pointed out earlier that the p+/n-well and n+/p-well junctions must remain in a reverse biasing condition all the time to ensure the existence of a junction capacitance. This imposes a stringent constraint on the swing of the voltage across the terminals of the varactors

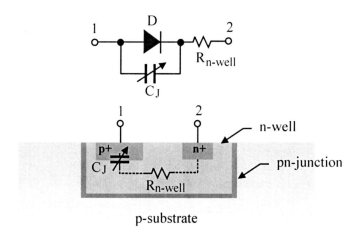

Figure 2.7.   Sideview of p+/n-well varactors

The sideview of accumulation-mode MOS varactors is shown in Fig.2.9 for $V_G < V_S$ and Fig.2.10 for $V_G > V_S$. If $V_G < V_S$, the electrons in the n-well region underneath the gate will be repelled and a depletion region is created. When $V_G > V_S$, the electrons from the n+ diffusion regions will be pulled to the region underneath the gate, creating an accumulation layer and C_{GS} arises to the gate-oxide capacitance. As pointed out in [68] that $\dfrac{C_{max}}{C_{min}}$ can be made from 2.5 to 3 when $-1V \leq v_{GS} \leq 1V$. A key advantage of accumulation-mode MOS varactors is the large voltage swing across the terminals of the varactors. They are the most widely used varactors in voltage/current-controlled oscillators.

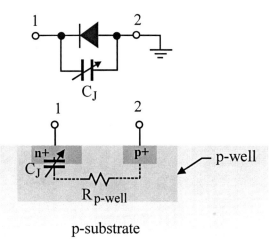

Figure 2.8. Sideview of n+/p-well varactors.

A common drawback of junction varactors and MOS varactors is their small capacitance tuning range.

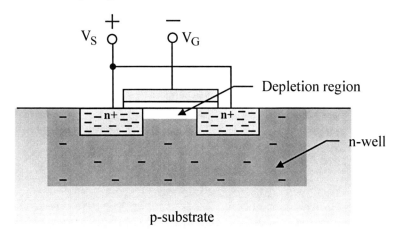

Figure 2.9. Sideview of accumulation-mode MOS varactors when $V_G < V_S$.

Conductance tuning can be done by varying the dc operating point of the transconductors. This approach offers a large conductance tuning range, subsequently a large inductance tuning range. The conductance tuning range is set by the constraint that the transconducting transistors of the transconductors must remain in the saturation. Conductance tuning can be used for the coarse tuning of the inductance while capacitance tuning can be used for the fine tuning of the inductance, as shown in Fig.2.12. The conductance of either the transconductor

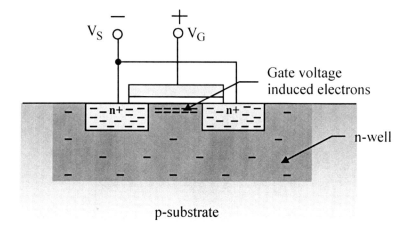

Figure 2.10.   Sideview of accumulation-mode MOS varactors when $V_G > V_S$.

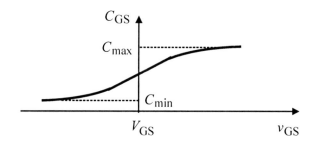

Figure 2.11.   Capacitance of accumulation-mode MOS varactors.

with a positive transconductance or that with a negative transconductance can be tuned. The conductance tuning range is set by the pinch-off condition while the capacitance tuning range is set by the range of the control voltage of the varactors.

It is seen from (2.9) that a change in the transconductances of the transconductors of an active inductor will not affect R_p and C_p of gyrator-C active inductors. It will, however, alter the parasitic series resistance R_s of the active inductor. This is echoed with a change in the quality factor of the active inductors. The variation of the quality factor due to the tuning of L must therefore be compensated for such that L and Q are tuned in a truly independent fashion. It should be noted that the fine tuning of the inductance of active inductors from the capacitance tuning does not affect R_s.

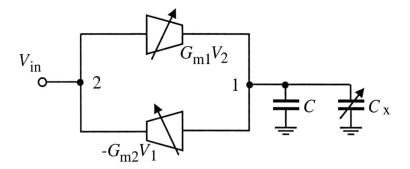

Figure 2.12.   Inductance tuning of gyrator-C active inductors. Conductance tuning can be carried out by varying either G_{m1} or G_{m2} while capacitance tuning is done by varying the varactor C_x.

2.2.3   Quality Factor

The quality factor Q of an inductor quantifies the ratio of the net magnetic energy stored in the inductor to its ohmic loss in one oscillation cycle. For spiral inductors, the quality factor of these inductors is independent of the voltage / current of the inductors. This property, however, does not hold for active inductors as the inductance of these inductors depends upon the transconductances of the transconductors constituting the active inductors and the load capacitance. When active inductors are used in applications such as LC oscillators, the inductance of the active inductors is a strong function of the swing of the voltage and current of the oscillators. To quantify the ratio of the net magnetic energy stored in the inductor to its ohmic loss in one oscillation cycle and relate it to the performance of LC oscillators, in particular, the phase noise of the oscillators, an alternative definition of the quality factor that accounts for the swing of the voltage / current of the active inductors is needed.

Instantaneous Quality Factor

The quality factor Q of an inductor quantifies the ratio of the net magnetic energy stored in the inductor to its ohmic loss in one oscillation cycle [33, 29]

$$Q \; = \; 2\pi \times \frac{\text{Net magnetic energy stored}}{\text{Energy dissipated in one oscillation cycle}}. \qquad (2.21)$$

For a linear inductor, the complex power of the active inductor is obtained from

$$P(j\omega) = I(j\omega)V^*(j\omega) = \Re e[Z]|I(j\omega)|^2 + j\Im m[Z]|I(j\omega)|^2, \qquad (2.22)$$

where $\Re e[Z]$ and $\Im m[Z]$ are the resistance and inductive reactance of the inductor, respectively, $V(j\omega)$ and $I(j\omega)$ are the voltage across and the current through the inductor, respectively, the superscript $*$ is the complex conjugation operator, and $|.|$ is the absolute value operator. The first term in (2.22) quantifies the net energy loss arising from the parasitic resistances of the inductor, whereas the second term measures the magnetic energy stored in the inductor. Eq.(2.21) in this case becomes

$$Q = \frac{\Im m[Z]}{\Re e[Z]}. \tag{2.23}$$

Eq.(2.23) provides a convenient way to quantify Q of linear inductors including active inductors.

Active inductors are linear when the swing of the voltages / currents of the inductors are small and all transistors of the active inductors are properly biased. The quality factor of a lossy gyrator-C active inductor can be derived directly from (2.15) and (2.23)

$$Q = \left(\frac{\omega L}{R_s}\right) \frac{R_p}{R_p + R_s\left[1 + \left(\frac{\omega L}{R_s}\right)^2\right]} \left[1 - \frac{R_s^2 C_p}{L} - \omega^2 L C_p\right]. \tag{2.24}$$

Fig.2.13 shows the frequency dependence of the quality factor of the active inductor with $R_s = 4\Omega$, $R_p = 1k\Omega$, $C_p = 140$ fF, and $L = 1.6$ nH [33]. It is seen that the first term in (2.24), denoted by

$$Q_1 = \frac{\omega L}{R_s}, \tag{2.25}$$

quantifies the quality factor of the active inductor at low frequencies. The second term, denoted by

$$Q_2 = \frac{R_p}{R_p + R_s\left[1 + \left(\frac{\omega L}{R_s}\right)^2\right]}, \tag{2.26}$$

accounts for the effect of the finite output impedance of deep sub-micron MOSFETs, whereas the third term, denoted by

$$Q_3 = 1 - \frac{R_s^2 C_p}{L} - \omega^2 L C_p, \tag{2.27}$$

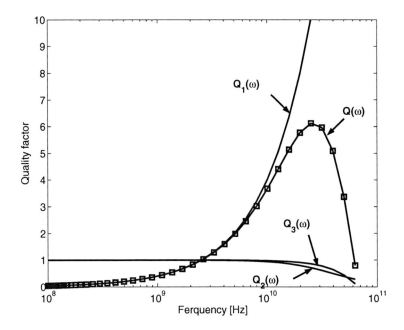

Figure 2.13. Frequency dependence of the quality factor of active inductors.

shows that the quality factor vanishes when frequency approaches the cut-off frequency of the transconductors of the active inductor. Q_2 and Q_3 manifest themselves at high frequencies only.

The sensitivity of the quality factor of the active inductor with respect to R_s and R_p is investigated in Figs.2.14 and 2.15, respectively. It is seen that Q_1 dominates the quality of the active inductor and is therefore widely used to quantify the quality factor of active inductors.

To boost the quality factor of active inductors, R_s must be minimized. Four approaches can be used to reduce R_s:

- *Approach 1* - Because $R_s = \dfrac{G_{o1}}{G_{m1}G_{m2}}$, R_s can be lowered by reducing G_{o1} directly. Since G_{o1} is typically the output impedance of the transconductor with a positive transconductance, the use of transconductors with a large output impedance is critical. As an example, consider the transconductors shown in Fig.2.16. The transconductance of the transconductor in Fig.2.16(a) is positive. This is because an increase in v_{in} will lead to an increase in v_{GS}, which in turn increases i_D. Because $i_o = i_D - J$, i_o will increase as well. The transconductance of the transconductor in Fig.2.16(b) is also positive. This is because an increase in v_{in} will decrease i_{D1}. As a result, $i_{D2} = J_1 - i_{D1}$ will increase. Since $i_o = i_{D2} - J_2$, i_o will increase as well. Although both transconductors have a positive transcon-

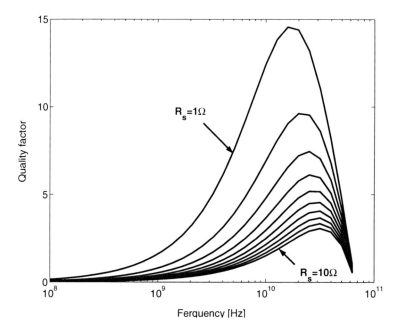

Figure 2.14. The effect of R_s on the quality factor of active inductors. R_s is varied from 1Ω to 10Ω with step 1Ω.

ductance and both have an infinite input impedance, the output impedance of the transconductor in Fig.2.16(a) is given by $\dfrac{1}{g_m}$ approximately whereas that of the transconductor in Fig.2.16(b) is given by r_{o2}. Active inductors constructed using the transconductor in Fig.2.16(b) will have a smaller R_s, subsequently a higher Q.

- *Approach 2* - R_s can be lowered by increasing G_{m1} and G_{m2} directly. Since the transconductances of the transconductors are directly proportional to the dc biasing currents and the width of the transistors of the transconductors, R_s can be lowered by either increasing the dc biasing currents or increasing the transistor width. The former, however, increases the static power consumption of the active inductors whereas the latter lowers the self-resonant frequency of the active inductors. Another downside of this approach is that the inductance of the inductors L will also be affected.

- *Approach 3* - Reduce G_{o1} using advanced circuit techniques, such as cascodes. Cascodes are effective in lowering the output conductance and can be used here to reduce G_{o1}, as shown in Fig.2.17. Table 2.1 compares the minimum supply voltage and output conductance of basic, cascode, regulated cascode, and multi-regulated cascode transconductors.

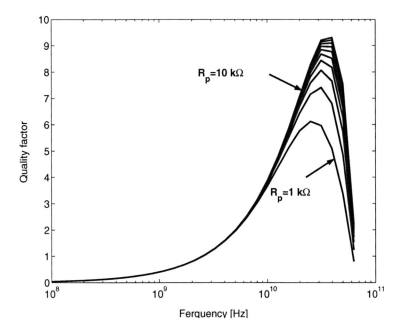

Figure 2.15. The effect of R_p on the quality factor of active inductors. R_p is varied from $1k\Omega$ to $10k\Omega$ with step $1k\Omega$.

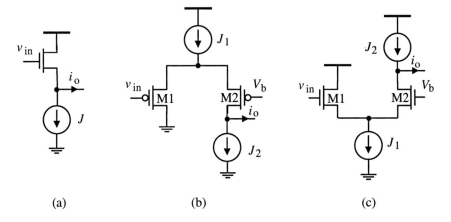

(a)   (b)   (c)

Figure 2.16. Simplified schematics of transconductors with a positive transconductor. (a) Common-drain transconductor; (b,c) Differential-pair transconductors.

- *Approach 4* - Use a shunt negative resistor at the output of the positive transconductor to cancel out the parasitic resistances, both series and parallel, of active inductors. It is well known that the series RL network of the RLC network shown in Fig.2.18 can be replaced with the parallel RL network shown in the figure with [68]

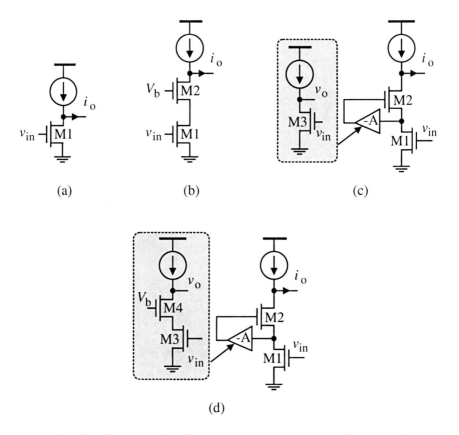

Figure 2.17. Simplified schematics of cascode transconductors. (a) Basic transconductor; (b) Cascode transconductor; (c) Regulated cascode transconductor; (d) Multi-regulated cascode transconductor.

$$R_p = R_s(1 + Q^2)$$

$$L_p = L_s\left(1 + \frac{1}{Q^2}\right), \qquad\qquad (2.28)$$

where $Q = \dfrac{\omega L_s}{R_s}$. This is because in order to have the two network to be equivalent, in other words, to exhibit the same terminal impedance, $Z_s(j\omega) = Z_p(j\omega)$ is required.

$$R_s + j\omega L_s = \frac{j\omega R_p L_p}{R_p + j\omega L_p}. \qquad\qquad (2.29)$$

Matching the real and imaginary parts yield

Table 2.1. The minimum supply voltage and output conductance of basic and cascode transconductors.

Transconductor	Min. V_{DD}	Output impedance
Basic	$2V_{sat}$	$G_o = \dfrac{1}{r_o}$
Cascode	$3V_{sat}$	$G_o = \dfrac{1}{r_{o1}(r_{o2}g_{m2})}$
Regulated cascode	$2V_T + V_{sat}$	$G_o = \dfrac{1}{r_{o1}(r_{o2}g_{m2})(r_{o3}g_{m3})}$
Multi-regulated cascode	$2V_T + V_{sat}$	$G_o = \dfrac{1}{r_{o1}(r_{o2}g_{m2})(r_{o3}g_{m3})(r_{o4}g_{m4})}.$

$$R_s R_p - \omega^2 L_s L_p = 0,$$
$$R_s L_p + R_p L_s = R_p L_p. \tag{2.30}$$

$$Q = \frac{\omega L_s}{R_s}$$

Eq.(2.28) follows from Eq.(2.30). Note that (2.28) is valid at all frequencies.

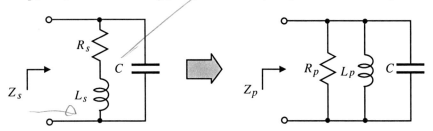

Figure 2.18. Transformation of a RL parallel branch to a RL series branch.

Now consider the RLC network of an active inductor shown in Fig.2.19. The $L \sim R_s$ branch of the RLC circuit of the active inductor is replaced with the parallel $\hat{L} \sim \hat{R}_p$ network with

$$\hat{L} = L\left(1 + \frac{1}{Q^2}\right),$$
$$\hat{R}_p = R_s(1 + Q^2). \tag{2.31}$$

Consider two cases :

- Case 1 - If R_s is negligible, the quality factor of the active inductor is mainly determined by R_p. From (2.24) with $R_s = 0$, we arrive at

$$Q = \frac{R_p}{\omega L}\left[1 - \left(\frac{\omega}{\omega_o}\right)^2\right]. \qquad (2.32)$$

At frequencies below ω_o, $\left(\dfrac{\omega}{\omega_o}\right)^2 \approx 0$ and

$$Q \approx \frac{R_p}{\omega L_s} \qquad (2.33)$$

follows.

- Case 2 - If R_p is large, the quality factor is mainly determined by R_s

$$Q = \frac{\omega L}{R_s}\left[1 - \left(\frac{\omega_z}{\omega_o}\right)^2 - \left(\frac{\omega}{\omega_o}\right)^2\right]. \qquad (2.34)$$

At frequencies above ω_z and below ω_o, $\left(\dfrac{\omega_z}{\omega_o}\right)^2 \approx 0$ and $\left(\dfrac{\omega}{\omega_o}\right)^2 \approx 0$. Eq.(2.34) is simplified to

$$Q \approx \frac{\omega L}{R_s}. \qquad (2.35)$$

As shown in Fig.2.19, the total parasitic parallel resistance of the active inductor becomes

$$R_{total} = R_p // \hat{R}_p. \qquad (2.36)$$

In this case, a negative resistor of resistance $R_{comp} = -R_{total}$ can be connected in parallel with C_p to eliminate the effect of both R_p and R_s of the active inductor simultaneously. Note that the resistance of the negative resistor should be made tunable such that a total cancellation can be achieved. The quality factor of the compensated active inductor at ω_o is given by

$$Q(\omega_o) = (R_p||\hat{R}_p||R_{comp})\sqrt{\frac{C_p + C_{comp}}{\hat{L}}}, \qquad (2.37)$$

where C_{comp} is the input capacitance of the compensating nega

It should be noted that because R_s and R_p are frequency-depen
should be designed in such a way that a total resistance ca
achieved across the frequency range of the active inductor. It sł
noted that although the negative resistor compensation technic
used to improve the quality factor of spiral inductors, a total con
this case is difficult to achieve. This is because an active negati
used to cancel out the largely skin-effect induced parasitic series resistance
of spirals.

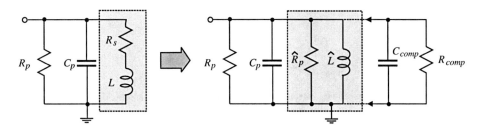

Figure 2.19. Q enhancement using a shunt negative resistor.

Average Quality Factors

Active inductors are RLC tanks when R_s, R_p, and C_p are accounted for.
The quality factor of LC tanks is obtained from [69]

$$Q(\omega) = \frac{\omega_o}{2} \frac{\partial \phi(\omega)}{\partial \omega},$$ (2.38)

where $\phi(\omega)$ is the phase of the tank impedance. The quality factor of a passive
LC tank at a given frequency is independent of the current of the tank.

Unlike passive LC tanks, the inductance of the active inductors varies with
the current / voltage of the inductors. The effective quality factor defined as

$$\overline{Q(\omega)} = \frac{\omega}{2(I_{max} - I_{min})} \int_{I_{min}}^{I_{max}} Q(\omega, i) di,$$ (2.39)

where I_{min} and I_{max} are the minimum and maximum currents of the transconductors of active inductors, and $Q(\omega_o, i)$ is the instantaneous quality factor
at frequency ω and channel current i provides an effective mean to quantify
the quality factor of active inductors, especially when active inductors are employed in circuits that are operated in a large-signal mode, such as LC tank
oscillators.

2.2.4   Noise

Active inductors exhibit a high level of noise as compared with their spiral counterparts. To analyze the noise of a gyrator-C active inductor, the power of the input-referred noise-voltage and that of the noise-current generators of the transconductors constituting the active inductor must be derived first. Fig.2.20 shows the partial schematics of basic transconductors widely used in the construction of gyrator-C active inductors. The power of the input-referred noise-voltage generator, denoted by $\overline{v_n^2}$, and that of the input-referred noise-current generator, denoted by $\overline{i_n^2}$, of these transconductors can be derived using conventional noise analysis approaches for 2-port networks [70], and the results are given in Table 2.2 where

$$\overline{i_{nD}^2} = 4kT(\gamma + R_g g_m) g_m \Delta f \tag{2.40}$$

represents the sum of the power of the thermal noise generated in the channel of MOSFETs and the thermal noise of the gate series resistance of MOSFETs, R_g is the gate series resistance, $\gamma = 2.5$ for deep sub-micron devices, T is the temperature in degrees Kelvin, and k is Boltzmann constant. The effect of the flicker noise of MOSFETs, which has a typical corner frequency of a few MHz [71], is neglected. The thermal noise of other parasitics of MOSFETs, such as the thermal noise of the bulk resistance of the source and drain diffusions, is also neglected.

To illustrate how the results of Table 2.2 are derived, consider the common-gate transconductor. To derive the input-referred noise-voltage generator $\overline{i_n^2}$ of the transconductor, we first short-circuit the input of the transconductor, as shown in Fig.2.21.

The output noise power of the transconductor due to i_{nD} is calculated

$$\overline{v_{no}^2} = r_o^2 \overline{i_{nD}^2}, \tag{2.41}$$

where r_o is the output resistance of the transistor. We then remove $\overline{i_{nD}^2}$ and apply v_n at the input of the transconductor, as shown in Fig.2.22. The output noise power of the transconductor is obtained

$$\overline{v_{no}^2} = (1 + g_m r_o)^2 \overline{v_n^2}. \tag{2.42}$$

Equating (2.41) and (2.42) yields

$$\overline{v_n^2} = \frac{r_o^2}{(1 + g_m r_o)^2} \overline{i_{nD}^2} \approx \frac{1}{g_m^2} \overline{i_{nD}^2}. \tag{2.43}$$

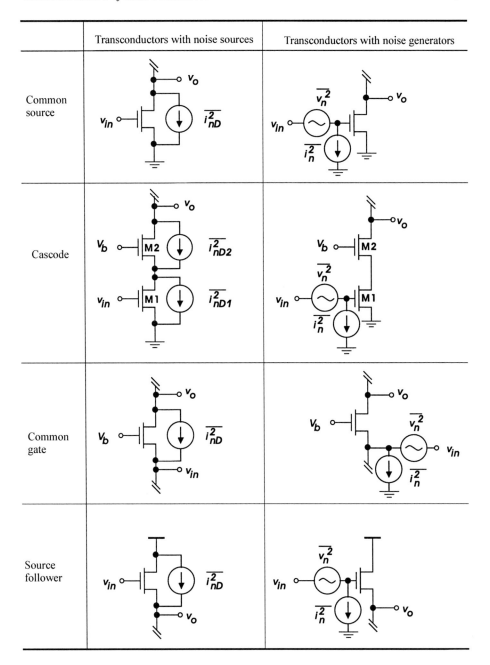

Figure 2.20. Input-referred noise-voltage and noise-current generators of transconductors at low frequencies.

Table 2.2.   Power of input-referred noise-voltage and noise-current generators of basic transconductors at low frequencies.

Transconductor	$\overline{v_n^2}$	$\overline{i_n^2}$
Common source	$\overline{v_n^2} = \dfrac{\overline{i_{nD}^2}}{g_m^2}$	$\overline{i_n^2} = 0$
Cascode	$\overline{v_n^2} = \dfrac{\overline{i_{nD1}^2}}{g_{m1}^2} + \dfrac{\overline{i_{nD2}^2}}{(g_{m1}r_{o1}g_{m2})^2}$	$\overline{i_n^2} = 0$
Common gate	$\overline{v_n^2} = \dfrac{\overline{i_{nD}^2}}{g_m^2}$	$\overline{i_n^2} = 0$
Source follower	$\overline{v_n^2} = \dfrac{\overline{i_{nD}^2}}{g_m^2}$	$\overline{i_n^2} = 0$

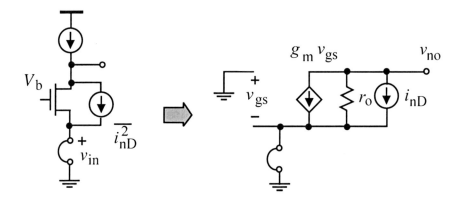

Figure 2.21.   Derivation of input-referred noise-voltage generator of a common-gate transconductor at low frequencies.

Note that we have utilized $r_o g_m \gg 1$ in (2.43) to simplify the results.

To derive the noise-current generator of the common-gate transconductor, consider Fig.2.23 where the input port of the transconductor is open-circuited and the output noise power of the circuit is calculated. To avoid the difficulty caused by the floating node 1, we assume that there exists a resistor of resistance

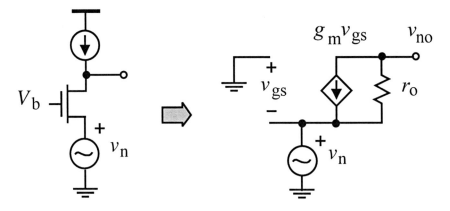

Figure 2.22. Derivation of the input-referred noise-voltage generator of a common-gate transconductor at low frequencies.

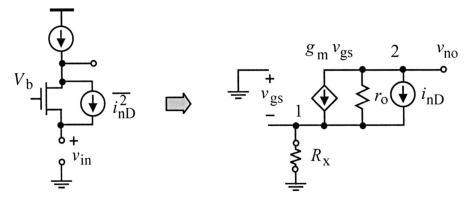

Figure 2.23. Derivation of the input-referred noise-current generator of a common-gate transconductor at low frequencies.

R_x between node 1 and the ground. Note that this approach is used in most IC CAD systems to avoid floating nodes. Writing KCL at nodes 1 and 2 yields

$$(g_x + g_m + g_o)v_1 - g_o v_2 = i_{nD} \quad \text{(node 1)},$$
$$-(g_o + g_m)v_1 + g_o v_2 + i_{nD} = 0 \quad \text{(node 2).}$$

(2.44)

Solving (2.44) yields

$$\overline{v_2^2} = \frac{1}{g_o^2} \overline{i_{nD}^2}.$$

(2.45)

We then apply the noise-current generator i_n at the input of the circuit, remove all the noise sources of the circuit, as shown in Fig.2.24, and compute the output noise power. Writing KCL at nodes 1 and 2 yields

$$(g_x + g_m + g_o)v_1 - g_o v_2 + i_n = 0 \quad \text{(node 1)},$$
$$-(g_o + g_m)v_1 + g_o v_2 = 0 \qquad\qquad \text{(node 2)}. \tag{2.46}$$

Solving (2.46) yields

$$\overline{v_2^2} = \left(1 + \frac{g_m}{g_o}\right)^2 \frac{1}{g_x^2} \overline{i_n^2}. \tag{2.47}$$

Equating (2.45) and (2.47) yields

$$\overline{i_n^2} = \left(\frac{g_x}{g_o}\right)^2 \frac{\overline{i_{nD}^2}}{\left(1 + \dfrac{g_m}{g_o}\right)^2}. \tag{2.48}$$

Taking the limit $R_x \to \infty$ or equivalently $g_x \to 0$ in (2.48), we arrive at

$$\overline{i_n^2} = 0. \tag{2.49}$$

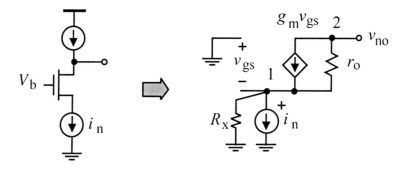

Figure 2.24.   Derivation of the input-referred noise-current generator of a common-gate transconductor at low frequencies.

Once $\overline{v_n^2}$ and $\overline{i_n^2}$ of the transconductors are available, the power of the input-referred noise-voltage and noise-current generators of active inductors can be derived.

Consider the active inductor of Fig.2.25(a) where $\overline{v_{n1}^2}$ and $\overline{v_{n2}^2}$ denote the power of the noise-voltage generators of the transconductors 1 and 2, respectively, and Y_1 and Y_2 are the admittance at ports 1 and 2, respectively. For the network of Fig.2.25(a), it is trivial to show that

$$\overline{V_1^2} = \overline{\left(V_{n1} + Y_1 \frac{G_{m2}V_{n2} + Y_2V_{n1}}{Y_1Y_2 + G_{m1}G_{m2}}\right)^2}. \tag{2.50}$$

For the network of Fig.2.25(b), we have

$$\overline{V_1^2} = \overline{\left(V_n + \frac{Y_1}{Y_1Y_2 + G_{m1}G_{m2}}I_n\right)^2}. \tag{2.51}$$

To ensure that Fig.2.25(a) and Fig.2.25(b) are equivalent, the right hand-side of (2.50) and that of (2.51) must be the same. To achieve this, we impose

$$V_n = V_{n1},$$
$$I_n = Y_2V_{n1} + G_{m2}V_{n2}. \tag{2.52}$$

Because

$$Z_{in}(s) = \frac{Y_1}{Y_1Y_2 + G_{m1}G_{m2}}, \tag{2.53}$$

we arrive at

$$V_n = V_{n1} + Z_{in}I_n. \tag{2.54}$$

For lossy gyrator-C active inductors, we have

$$Y_1 = G_{o1} + sC_1,$$
$$Y_2 = G_{o2} + sC_2, \tag{2.55}$$

Eq.(2.52) becomes

$$V_n = V_{n1},$$
$$I_n = (G_{o2} + sC_2)V_{n1} + G_{m2}V_{n2}. \tag{2.56}$$

By assuming V_{n1} and V_{n2} are uncorrelated, we arrive at

$$\overline{V_n^2} = \overline{V_{n1}^2},$$

$$\overline{I_n^2} = \left| G_{o2} + j\omega C_2 \right|^2 \overline{V_{n1}^2} + G_{m2}^2 \overline{V_{n2}^2}.$$

(2.57)

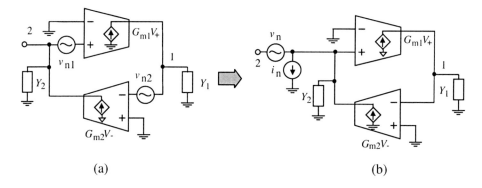

(a)   (b)

Figure 2.25.   Noise of single-ended gyrator-C active inductors.

2.2.5   Linearity

The preceding development of gyrator-C active inductors assumes that the transconductors of the active inductors are linear. This assumption is only valid if the swing of the input voltage of the transconductors is small. When the voltage swing is large, the transconductors will exhibit a nonlinear characteristic and the synthesized active inductors are no longer linear. The linearity constraint of active inductors sets the maximum swing of the voltage of the active inductors. If we assume that the transconductances of the transistors of gyrator-C active inductors are constant when the transistors are biased in the saturation, then the maximum swing of the voltage of the active inductors can be estimated from the pinch-off condition of the transistors. When the transistors of active inductors enter the triode region, the transconductances of the transistors decrease from g_m (saturation) to g_{ds} (triode) in a nonlinear fashion, as illustrated graphically in Fig.2.26. It should be emphasized that although the transconductances of the transconductors of gyrator-C networks drop when the operating point of the transistors of the transconductors moves from the saturation region to the triode region, the inductive characteristics at port 2 of the gyrator-C network remain. The inductance of the gyrator-C active inductors, however, increases from $L = \dfrac{C}{G_{m1}G_{m2}}$ to $L = \dfrac{C}{G_{ds1}G_{ds2}}$, where $G_{m1,m2}$ are the transconductances of transconductors 1 and 2 respectively when in the saturation and $G_{ds1,ds2}$ are the transconductances of transconductors 1 and 2 respectively when in the triode. Note that we have assumed the load capacitance C remains unchanged. The inductance thus varies with the swing of the voltage of the active inductors in a nonlinear fashion. It should also be noted that the parasitic resistances of active inductors also vary with the voltage swing of the active inductors.

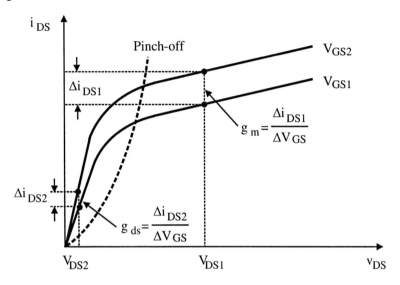

Figure 2.26. Transconductance of MOSFETs in the saturation and triode regions. Because $\Delta i_{DS1} > \Delta i_{DS2}$, $g_m > g_{ds}$ follows.

2.2.6 Stability

Gyrator-C active inductors are negative feedback systems. The stability of active inductors is critical to the overall stability of systems employing active inductors. In this section, we investigate the stability of gyrator-C active inductors.

The impedance looking into port 2 of the gyrator-C active inductor shown in Fig.2.4 is given by

$$Z = \frac{sC_1 + G_{o1}}{s^2 C_1 C_2 + s(C_1 + C_2) + G_{m1}G_{m2}}, \tag{2.58}$$

where we have utilized $G_m \gg G_o$ to simply the results. The poles of the system are given by

$$p_{1,2} = \frac{C_1 + C_2}{2C_1 C_2}\left[-1 \pm \sqrt{1 - \frac{4C_1 C_2 G_{m1} G_{m2}}{(C_1 + C_2)^2}}\right]. \tag{2.59}$$

The poles of the gyrator-C active inductor are located in the left half of the
s-plane and the gyrator-C active inductor is a stable system.

The degree of stability can be assessed by evaluating its damping factor,
which is obtained by comparing the denominator of (2.58) with the standard
form of the characteristic equation of second-order systems

$$s^2 + 2\omega_o\xi s + \omega_o^2 = 0, \tag{2.60}$$

where ξ denotes the damping factor and ω_o is the pole resonant frequency. The
result is given by

$$\xi = \frac{1}{2\sqrt{G_{m1}G_{m2}}}\left(\sqrt{\frac{C_2}{C_1}} + \sqrt{\frac{C_1}{C_2}}\right). \tag{2.61}$$

Eq.(2.61) reveals that an increase in G_{m1} and G_{m2} will lead to a decrease in ξ.
This is echoed with an increase in the level of oscillation in the response of the
active inductor. Also observed from (2.61) is that the ratios $\dfrac{C_1}{C_2}$ and $\dfrac{C_2}{C_1}$ have
a marginal impact on the damping factor simply because these two quantities
vary in the opposite directions when C_1 and C_2 change, and the values of C_1
and C_2 are often close.

If $C_1 = C_2 = C$ and $G_{m1} = G_{m2} = G_m$, we have

$$p_{1,2} = \frac{1}{C}\left(-1\pm\sqrt{1 - G_m^2}\right),$$

$$\xi = \frac{1}{G_m}. \tag{2.62}$$

An increase of G_m will lead to a decrease of ξ. This is echoed with a reduced
level of damping. Because $\Re e[p_{1,2}] = -\dfrac{1}{C}$, the absolute stability margin is
set by the capacitance C and is independent of G_m. It should be noted that
the preceding analysis is based on the assumption that active inductors are
2nd-order systems. When the parasitics of MOSFETs are accounted for, active
inductors are no longer 2nd-order systems and their stability will deteriorate.

2.2.7 Supply Voltage Sensitivity

The supply voltage sensitivity of the inductance of active inductors is a
figure-of-merit quantifying the effect of the variation of the supply voltage on
the inductance of the active inductors. The fluctuation of the supply voltage of a
mixed analog-digital system is mainly due to the switching noise of the system

[14]. Assume that the supply voltage of a mixed-mode system containing an active inductor varies from V_{DD} and $V_{DD} + \Delta V_{DD}$, where ΔV_{DD} is a random variable with $E[\Delta V_{DD}] = 0$, where $E[.]$ denotes the mathematical mean operator. For a well designed mixed-mode system, $\Delta V_{DD} \ll V_{DD}$ holds. The small-signal analysis approach can therefore be employed to analyze the effect of ΔV_{DD} on the inductance of the active inductor. Following the definition of normalized sensitivity given in [72], the normalized sensitivity of the inductance of an active inductor to the supply voltage is defined as

$$S_{V_{DD}}^{L} = \frac{V_{DD}}{L} \frac{\partial L}{\partial V_{DD}}. \tag{2.63}$$

The fluctuation of the supply voltage V_{DD} affects the inductance of the active inductor mainly by altering the dc operating point, subsequently the transconductances of the transconductors constituting the active inductor. By assuming that the load capacitance C of the gyrator-C active inductor does not vary with V_{DD} and because $L = \dfrac{C}{G_{m1} G_{m2}}$, we arrive at

$$\frac{\partial L}{\partial V_{DD}} = -L \left(\frac{1}{G_{m1}} \frac{\partial L}{\partial G_{m1}} + \frac{1}{G_{m2}} \frac{\partial L}{\partial G_{m2}} \right). \tag{2.64}$$

The normalized supply voltage sensitivity of the active inductor is given by

$$S_{V_{DD}}^{L} = -\left(S_{V_{DD}}^{G_{m1}} + S_{V_{DD}}^{G_{m2}} \right), \tag{2.65}$$

where $S_{V_{DD}}^{G_{m1}}$ and $S_{V_{DD}}^{G_{m2}}$ are the normalized supply voltage sensitivity of G_{m1} and G_{m2}, respectively. Eq.(2.65) reveals that both $S_{V_{DD}}^{G_{m1}}$ and $S_{V_{DD}}^{G_{m2}}$ contribute equally to $S_{V_{DD}}^{L}$. To minimize the supply voltage sensitivity of active inductors, transconductors with a constant G_m should be used.

2.2.8 Parameter Sensitivity

The minimum feature size of MOS devices in modern CMOS technologies has been scaled down more aggressively as compared with the improvement in process tolerance such that the effect of process variation on the characteristics of circuits becomes increasingly critical. For example, the resistance of poly resistors in a typical $0.18\mu m$ CMOS process has an error of $\pm 20\%$ approximately and that of n-well resistors has an error of $\pm 30\%$ approximately. Analysis of the effect of parameter spread is vital to ensure that the performance of circuits meets design specifications once the circuits are fabricated. Active inductors consist of a number of active devices and their performance is greatly affected by the parameter spread of these components. The normalized sensitivity of the inductance of an active inductor to a parameter x_j of the inductor defined as

$$S_{x_j}^L = \frac{x_j}{L} \frac{\partial L}{\partial x_j} \tag{2.66}$$

quantifies the effect of the variation of the parameter x_j on the inductance of the active inductor. By assuming that the parameters of the active inductor are Gaussian distributed and uncorrelated, the overall effect of the variation of the parameters of the active inductor on the inductance of the inductor is obtained from

$$\sigma_L^2 = \sum_{j=1}^{N} \left(\frac{\partial L}{\partial x_j} \right)^2 \sigma_{x_j}^2, \tag{2.67}$$

where σ_L and σ_{x_j} denote the standard deviations of L and x_j, respectively, and N is the number of the parameters of the active inductor. For a gyrator-C active inductor, because

$$\frac{\partial L}{C} = \frac{1}{G_{m1} G_{m2}},$$

$$\frac{\partial L}{G_{m1}} = -\frac{C}{G_{m1}^2 G_{m2}}, \tag{2.68}$$

$$\frac{\partial L}{G_{m2}} = -\frac{C}{G_{m1} G_{m2}^2},$$

we obtain the normalized spread of the inductance of the active inductor

$$\frac{\sigma_L^2}{L^2} = \frac{\sigma_C^2}{C^2} + \frac{\sigma_{G_{m1}}^2}{G_{m1}^2} + \frac{\sigma_{G_{m2}}^2}{G_{m2}^2}. \tag{2.69}$$

There are two ways in which circuit designers can analyze the effect of parameter spread on the inductance of active inductors, namely worst-case analysis, also known as corner analysis, and Monte Carlo analysis. The former determines the inductance of active inductors at process corners while the latter quantifies the degree of the spread of the inductance of active inductors around the nominal inductance of the inductors. The accuracy of Monte Carlo analysis increases with an increase in the number of simulation runs and is therefore extremely time consuming. Corner analysis, on the other hand, is time-efficient but the results obtained from corner analysis are typically over conservative. Despite of this, corner analysis is the most widely used method to quantify the effect of process spread.

2.2.9   Signal Sensitivity

Unlike spiral inductors whose inductance is independent of the voltage and current of the inductors, the inductance of gyrator-C active inductors varies with the voltage and current of the transconductors constituting the active inductors. This is because the transconductances G_{m1} and G_{m2} of the transconductors are signal dependent when signal swing is large. When an active inductor is used in applications where the voltage of the active inductor experiences a large degree of variation, such as active inductor LC oscillators, the transconductances of the transconductors of the active inductor vary with the signal swing. As a result, the inductance, parasitic resistances, and quality factor of the active inductor all vary with the signal swing.

2.2.10   Power Consumption

Spiral inductors do not consume static power. Gyrator-C active inductors, however, consume dc power, mainly due to the dc biasing currents of their transconductors. The power consumption of gyrator-C active inductors themselves is usually not of a critical concern because the inductance of these inductors is inversely proportional to the transconductances of the transconductors constituting the inductors. To have a large inductance, G_{m1} and G_{m2} are made small. This is typically achieved by lowering the dc biasing currents of the transconductors. When replica-biasing is used to minimize the effect of supply voltage fluctuation on the inductance of active inductors, as to be seen shortly, the power consumed by the replica-biasing network must be accounted for. Also, when negative resistors are employed for boosting the quality factor of active inductors, their power consumption must also be included. Often the power consumption of an active inductor is set by that of its replica-biasing and negative resistor networks.

2.3   Implementation of Single-Ended Active Inductors

The need for a high self-resonant frequency of active inductors requires that the transconductors of these active inductors be configured as simple as possible. This also lowers their level of power consumption and reduces the silicon area required to fabricate the inductors. Most reported gyrator-C active inductors employ a common-source configuration as negative transconductors, common-gate, source follower, and differential pair configurations as positive transconductors. These basic transconductors have the simplest configurations subsequently the highest cutoff frequencies and the lowest silicon consumption.

The load capacitor of the transconductors is realized using the intrinsic capacitance C_{gs} of the transistors of the transconductors directly to maximize the upper bound of the frequency range of the active inductors and to avoid the use of expensive floating capacitors, which are available only in mixed-mode CMOS processes. MOS varactors are often added in parallel with C_{gs} to tune the inductance of active inductors.

This section presents a comprehensive treatment of both the circuit implementation and characteristics of CMOS active inductors. To simplify analysis, the following assumptions are made in analysis of active inductors and in determination of both their signal swing and the minimum supply voltage : (i) nMOS and pMOS transistors have the same threshold voltage V_T. (ii) nMOS and pMOS transistors have the same pinch-off voltage V_{sat}. (iii) Only C_{gs} is considered. C_{gd} and parasitic diffusion capacitances are neglected unless otherwise noted explicitly. (iv) The minimum voltage drop across biasing current sources and current-source loads is V_{sat}.

2.3.1   Basic Gyrator-C Active Inductors

Fig.2.27 show the schematic of two basic gyrator-C active inductors. In Fig.2.27(a), the transconductor with a positive transconductance is common-gate configured while the transconductor with a negative transconductance is common-source configured. In Fig.2.27(b), the transconductor with a positive transconductance is common-drain configured while the transconductor with a negative transconductance is common-source configured. All transistors are biased in the saturation. A notable advantage of the active inductor in Fig.2.27(b) is that all transistors are nMOS, making it attractive for high-frequency applications.

For Fig.2.27(a), we have $C_1 = C_{gs2}$, $G_{o1} \approx g_{o1}$, $G_{m1} = g_{m1}$, $C_2 = C_{gs1}$, $G_{o2} \approx g_{m1}$, and $G_{m2} = g_{m2}$, where g_{oj} and g_{mj}, $j = 1, 2$, are the output conductance and transconductance of transistor j, respectively. Using (2.9), we obtain the parameters of the equivalent RLC network of the active inductor

$$C_p = C_{gs1},$$

$$R_p = \frac{1}{g_{m1}},$$

$$L = \frac{C_{gs2}}{g_{m1}g_{m2}},$$ (2.70)

$$R_s = \frac{g_{o1}}{g_{m1}g_{m2}}.$$

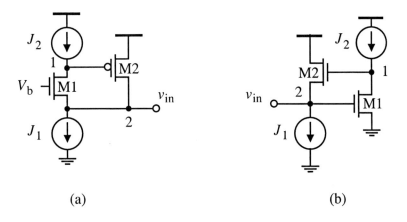

(a)   (b)

Figure 2.27. Simplified schematic of basic gyrator-C active inductors.

It is observed from (2.70) that the parasitic parallel resistance R_p is rather small, limiting the quality factor of the active inductor. Also, the parasitic series resistance is large, further lowering the quality factor. In evaluating the quality factor of this active inductor, the effect of the parasitic series resistance R_s is often neglected as R_p is small. In this case, the quality factor of the active inductor is obtained from

$$Q \approx \frac{R_p}{\omega L} = \frac{\omega_{t2}}{\omega}.$$ (2.71)

The self-resonant frequency of the active inductor is given by

$$\omega_o \approx \frac{1}{\sqrt{LC_p}} = \sqrt{\omega_{t1}\omega_{t2}},$$ (2.72)

where

$$\omega_{tj} = \frac{g_{mj}}{C_{gsj}}, i = 1, 2 \qquad (2.73)$$

is the cutoff frequency of transconductor j. At the self-resonance frequency of the active inductor $\omega_o = \sqrt{\omega_{t1}\omega_{t2}}$, the quality factor becomes

$$Q(\omega_o) = \sqrt{\frac{\omega_{t2}}{\omega_{t1}}}. \qquad (2.74)$$

The frequency of the zero of the active inductor, which is the lower bound of the frequency range of the active inductor, is given by (2.18)

$$\omega_z = \frac{g_{o1}}{C_{gs2}}. \qquad (2.75)$$

Eqs.(2.72) and (2.75) reveal that :

- In order to maximize the frequency range of the active inductor, ω_z should be minimized. This can be achieved by reducing g_{o1} or increasing C_{gs2}. The former is usually preferred as the latter lowers ω_o.

- Because the output impedance of deep sub-micron MOSFETs is small. The detrimental effect of $R_p = \dfrac{1}{g_{m1}}$ on the quality factor of the active inductor can not be neglected. The effect of R_p, however, can be eliminated by connecting a negative resistor of resistance $\hat{R}_p = -R_p$ in parallel with R_p.

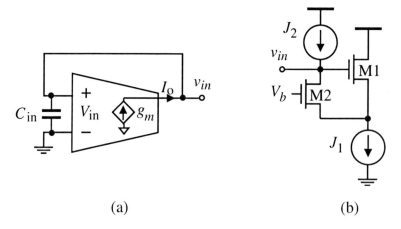

(a)   (b)

Figure 2.28. Single-ended negative impedance networks. (a) Block diagram. (b) Circuit implementation.

Negative resistors can be realized using transconductors with positive feedback, as shown in Fig.2.28 for single-ended negative resistors and Fig.2.29 for differential negative resistors. The positive feedback of the single-ended negative resistor in Fig.2.28 is depicted as the followings : An increase in the gate voltage of M_1 will increase the voltage at the source of M_1. Since M2 is a common-gate configuration, an increase in the source voltage of M2 will increase the drain voltage of M2. A positive feedback is thus established. Readers can verify the positive feedback of the differential negative resistors of Fig.2.29 in a similar manner. For Fig.2.29. It can be shown that the impedance at low frequencies is given by

$$Z \approx - \left(\frac{1}{g_{m1}} + \frac{1}{g_{m2}} \right). \tag{2.76}$$

To maximize the frequency range over which a constant negative resistance exists, transconductors synthesizing negative resistors should be configured as simple as possible.

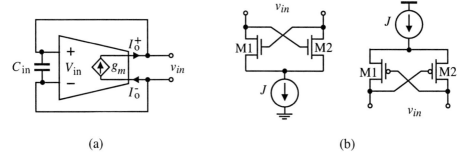

(a)   (b)

Figure 2.29. Differential negative impedance networks. (a) Block diagram. (b) Circuit implementation. The tail current source in the differential configuration can be removed, provided that biasing currents are provided by the circuit connected to the negative resistor. Note that the removal of the biasing tail current source will also remove the tunability of the resistance of the negative resistor.

▪ The effect of R_s can be compensated for in three different ways :

   – Use cascodes and regulated cascodes to reduce g_{o1} [37, 38, 65, 73]. It is well known that cascodes and regulated cascodes are effective means to boost the output impedance of transconductors. The price paid, however, is the reduced signal swing.

   – The reason that the preceding basic active inductors have a low R_p is because the input impedance of the positive transistor in Fig.2.27(a) is $1/g_{m1}$ and the output impedance of the positive transconductor in Fig.2.27(b) is $1/g_{m2}$. The use of transconductors that have both a large output impedance and a large input impedance will eliminate this drawback. As an example, the differentially-configured positive transconductor of Karsilayan-Schaumann active inductors shown in Fig.2.16(b,c) has an infinite input impedance and a large output impedance r_{o2} [74, 75, 48, 49, 61].

– Because the series RL branch of the RLC equivalent network of the active inductor can be replaced with a parallel $\hat{R}\hat{L}$ branch. The inductance of the parallel $\hat{R}\hat{L}$ branch is given by $\hat{L} = L\left(1 + \dfrac{1}{Q^2}\right)$ while the resistance is given by $\hat{R}_p = (1 + Q^2)R_s$. The total parasitic parallel resistance of the active inductor becomes $R_{p,total} = R_p \| \hat{R}_p$. The quality factor of the active inductor can be made infinite theoretically by employing a shunt negative resistor whose resistance is $-R_{p,total}$.

Table 2.3 compares the range of the voltage swing and the minimum supply voltage of the two basic active inductors. It is seen that the active inductor in Fig.2.27(a) offers a large input voltage swing and requires a lower minimum supply voltage.

Table 2.3. Comparison of the input voltage swing and the minimum supply voltage of the basic active inductors in Fig.2.27.

Active inductor	Fig.2.27(a)	Fig.2.27(b)
Max. input voltage	$V_{DD} - V_T - V_{sat}$	$V_{DD} - V_T - V_{sat}$
Min. input voltage	V_{sat}	V_T
Min. V_{DD}	$V_T + 2V_{sat}$	$2V_T + V_{sat}$

2.3.2   Wu Current-Reuse Active Inductors

Fig.2.30 show the schematic of Wu current-reuse active inductors proposed in [42, 45, 36]. In the nMOS version of the active inductor, the positive transconductor is common-gate configured while the negative transconductor is common-source configured. When only C_{gs} is considered, $C_1 = C_{gs2}$, $G_{o1} \approx g_{o1} + g_{o2}$, $G_{m1} = g_{m1}$, $C_2 = C_{gs1}$, $G_{o2} = \dfrac{1}{g_{m1}}$, and $G_{m2} = g_{m2}$. The parameters of the equivalent RLC network of this active inductor are given by

$$C_p = C_{gs1},$$

$$R_p = \frac{1}{g_{m1}},$$

$$L = \frac{C_{gs2}}{g_{m1}g_{m2}}, \tag{2.77}$$

$$R_s = \frac{g_{o1} + g_{o2}}{g_{m1}g_{m2}}.$$

The quality factor of Wu active inductors can be estimated by neglecting the effect of R_s and only focusing on R_p as R_p is small.

$$Q \approx \frac{R_p}{\omega L} = \frac{\omega_{t2}}{\omega}. \tag{2.78}$$

At the self-resonant frequency $\omega_o = \sqrt{\omega_{t1}\omega_{t2}}$, we have

$$Q(\omega_o) \approx \sqrt{\frac{\omega_{t2}}{\omega_{t1}}}. \tag{2.79}$$

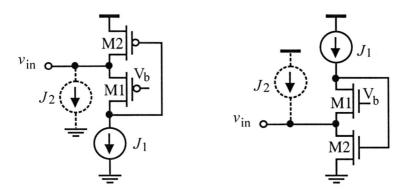

Figure 2.30. Simplified schematic of Wu current reuse active inductors.

It is seen from the preceding analysis that to increase ω_o, both ω_{t1} and ω_{t2} need to be increased. Increasing ω_{t1}, however, lowers $Q(\omega_o)$. Increasing ω_{t1} should therefore be avoided. To boost ω_{t2} without increasing ω_{t1}, the dc biasing current of M_1 is kept unchanged while that of M_2 is increased by injecting an additional current J_2 into M_2. The additional current source J_2 is used to boost the transconductance of M_2 such that the upper frequency bound can be increased without lowering the quality factor. In practical design, J_2 is provided by the stage preceding to the inductors and the active inductors are known as Wu current-reuse active inductors.

2.3.3   Lin-Payne Active Inductors

The Lin-Payne active inductor proposed in [39] and shown in Fig.2.31 requires the minimum supply voltage of only $V_T + 2V_{sat}$. Another implementation of Lin-Payne active inductors is shown in Fig.2.32 [76]. The minimum supply voltage of the active inductor is also $V_T + 2V_{sat}$. They can be analyzed in a similar way as Wu current re-use active inductors.

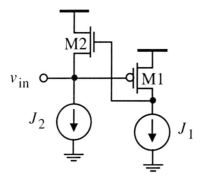

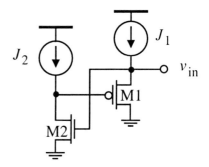

Figure 2.31.   Simplified schematic of Lin-Payne active inductor.

Figure 2.32.   Simplified schematic of the variation of Lin-Payne active inductor.

2.3.4   Ngow-Thanachayanont Active Inductors

Ngow and Thanachayanont proposed the low-voltage active inductor shown in Fig.2.33[76]. The addition of M_3 branch relaxes the biasing difficulties encountered in Lin-Payne active inductors. M_{1-3} can easily be biased in the saturation to ensure a stable operation of the active inductors. The analysis of Ngow-Thanachayanont active inductors is similar to that of Wu current re-use active inductors and is left as an exercise for readers.

2.3.5   Hara Active Inductors

Although the MESFET implementation of Hara active inductors appeared two decades ago [77, 78], the CMOS version of Hara active inductors only emerged a few years ago [41, 40, 79–82]. Hara active inductors shown in Fig.2.34 employ only a MOSFET and a resistor. They are indeed gyrator-C active inductors. The feedback operation of nMOS Hara active inductor is as the followings : An increase of the input current will result in an increase in the voltage at the input node. Since the gate voltage is kept at V_{DD}, v_{GS} is reduced. This in turn lowers the current flowing out of the active inductor.

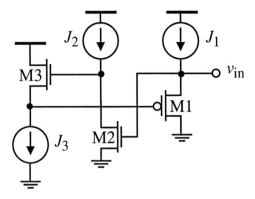

Figure 2.33. Simplified schematic of low-voltage gyrator-C active inductor.

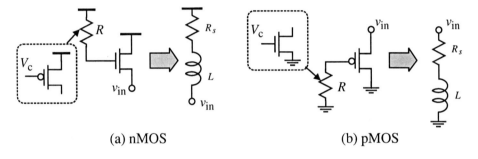

(a) nMOS   (b) pMOS

Figure 2.34. Hara active inductors. Resistor R can be made variable by implementing it using MOSFETs biased in the triode.

The input impedance of the nMOS Hara active inductor can be derived from its small-signal equivalent circuit shown in Fig.2.35

$$Z \approx \left(\frac{1}{RC_{gs}C_{gd}}\right) \frac{sRC_{gd} + 1}{s^2 + s\dfrac{g_m}{C_{gs}} + \dfrac{g_m}{RC_{gs}C_{gd}}}, \qquad (2.80)$$

where $g_m \gg g_o$ and $C_{gs} \gg C_{gd}$ were utilized to simplify the results.

The self-resonant frequency ω_o and the frequency of the zero ω_z of the active inductor are given by

$$\omega_o = \sqrt{\frac{g_m}{RC_{gs}C_{gd}}} = \sqrt{\omega_t \omega_z},$$

$$\omega_z = \frac{1}{RC_{gd}}, \qquad (2.81)$$

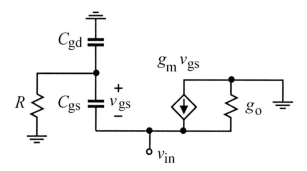

Figure 2.35.   Small-signal equivalent circuit of nMOS Hara active inductors.

where

$$\omega_t = \frac{g_m}{C_{gs}}. \tag{2.82}$$

The Bodé plots of Hara active inductors are the same as those given in Fig.2.6. The network exhibits an inductive characteristic in the frequency range $\omega_z < \omega < \omega_o$. When C_{gd}, C_{sb}, C_{sb}, and high-order effects are neglected, the inductance L and parasitic series resistance R_s of nMOS Hara active inductors are given by

$$R_s = \frac{g_m + \omega^2 C_{gs}^2 R}{g_m^2 + \omega^2 C_{gs}^2}, \tag{2.83}$$

and

$$L = \frac{C_{gs}(g_m R - 1)}{g_m^2 + \omega^2 C_{gs}^2}. \tag{2.84}$$

Observe that

$$g_m R > 1 \tag{2.85}$$

is required to ensure $L > 0$. Under the condition $g_m R \gg 1$, Eqs.(2.83) and (2.84) can be written as

$$R_s = \frac{\dfrac{1}{g_m} + \left(\dfrac{\omega}{\omega_t}\right)^2 R}{1 + \left(\dfrac{\omega}{\omega_t}\right)^2} \approx \frac{1}{g_m}, \tag{2.86}$$

and

$$L = \frac{R}{\omega_t \left[1 + \left(\dfrac{\omega}{\omega_t} \right)^2 \right]} \approx \frac{R}{\omega_t}, \tag{2.87}$$

where we have utilized $\left(\dfrac{\omega}{\omega_t} \right)^2 \approx 0$. It is evident from (2.87) that the inductance L is directly proportional to R and can be tuned by varying R. The series resistance R_s is largely dominated by g_m at frequencies below the cut-off frequency of the transistor.

The dependence of the input impedance of the nMOS Hara active inductor on the resistance of the resistor R and the width of the transistor is shown in Fig.2.36 and Fig.2.37, respectively. The active inductor was implemented in TSMC-0.18μm 1.8V CMOS technology and analyzed using SpectreRF with BSIM3V3 device models. It is observed that an increase of R will lower both ω_z and ω_o. This agrees with the theoretical results. Increasing the width of the transistor will lower ω_o because

$$\omega_t = \frac{g_m}{C_{gs}} \approx \frac{3}{2} \frac{\mu_n}{L_c^2} (V_{GS} - V_T), \tag{2.88}$$

where μ_n is the surface mobility of free electrons and L_c is the channel length, is independent of the width of the transistor. Also, L is nearly independent of g_m.

Hara active inductors suffer from the loss of the voltage headroom of at least V_T. In [41], a voltage doubler was used to increase the supply voltage for R. A drawback of this approach is the complexity of the voltage doubler and the need for a control clock.

2.3.6 Wu Folded Active Inductors

The drawback of Hara active inductors can be eliminated by employing Wu folded active inductors shown in Fig.2.38[83]. Wu folded active inductors were initially proposed by Thanachayanont in [84].

To derive the parameters of the RLC equivalent circuit of Wu folded active inductors, consider the nMOS version of Wu folded active inductors with its small-signal equivalent circuit shown in Fig.2.39. To simplify analysis, we neglect C_{gd}, g_o and other parasitic capacitances of the transistor. It can be shown that the input impedance is given by

$$Z = \frac{sRC_{gs} + 1}{sC_{gs} + g_m}. \tag{2.89}$$

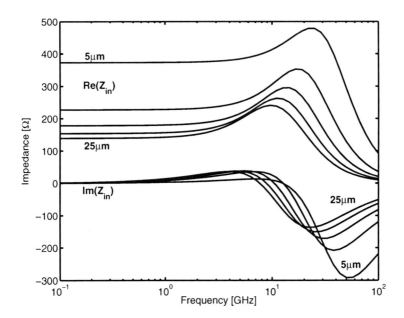

Figure 2.36. Dependence of Z of nMOS Hara active inductors on the width of the transistor. W is varied from $5\mu m$ to $25\mu m$ with step $5\mu m$, $R = 1k\Omega$. DC biasing current 0.5mA.

It becomes evident that Z has a zero at frequency $\omega_z = \dfrac{1}{RC_{gs}}$ and a pole at frequency $\omega_p = \dfrac{g_m}{C_{gs}}$. The network is resistive at low frequencies $\omega < \omega_z$ with resistance $R \approx \dfrac{1}{g_m}$ and inductive when $\omega_z < \omega < \omega_p$. Note the behavior of the network beyond ω_p can not be quantified by (2.89) due to the omission of C_{gd}.

To derive its RLC equivalent circuit, we examine the input admittance of the network

$$
\begin{aligned}
Y_{in} &= \frac{sC_{gs} + g_m}{sRC_{gs} + 1} \\
&= \frac{1}{R} + \frac{1}{s\dfrac{RC_{gs}}{g_m - \frac{1}{R}} + \dfrac{1}{g_m - \frac{1}{R}}}.
\end{aligned}
\tag{2.90}
$$

Eq.(2.90) can be represented by a series RL network in parallel with a resistor R_p with

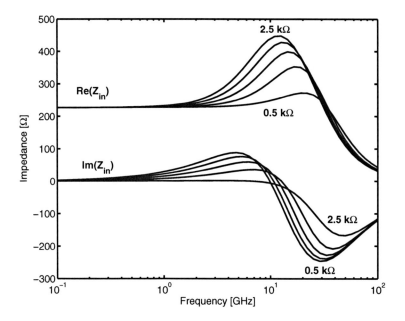

Figure 2.37. Dependence of Z of nMOS Hara active inductors on R. R is varied from 0.5 kΩ to 2.5 kΩ with step 0.5 kΩ, $W = 10\mu m$. DC biasing current 0.5mA.

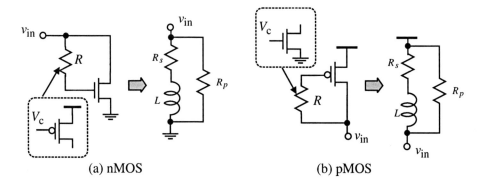

(a) nMOS   (b) pMOS

Figure 2.38. Simplified schematic of Wu folded active inductors.

$$R_p = R,$$

$$L = \frac{RC_{gs}}{g_m - \frac{1}{R}},$$

$$R_s = \frac{1}{g_m - \frac{1}{R}}.$$

(2.91)

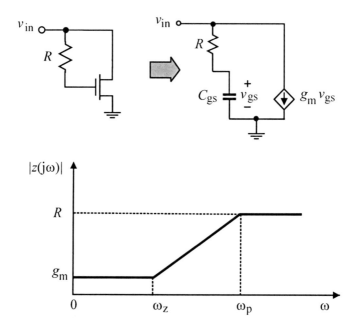

Figure 2.39. Small-signal equivalent circuit of the nMOS version of Wu folded active inductors.

It becomes evident that $g_m > \dfrac{1}{R}$ is required in order to have $L, R_s > 0$. Also, if $g_m \gg \dfrac{1}{R}$, we have $R_s \approx \dfrac{1}{g_m}$ and $L \approx \dfrac{RC_{gs}}{g_m} = \dfrac{R}{\omega_t}$. They are the same as those of the corresponding Hara active inductor investigated earlier.

2.3.7   Karsilayan-Schaumann Active Inductors

The active inductor proposed by Karsilayan and Schaumann is shown in Fig.2.40 [74, 75, 48, 49, 61]. It was also developed by Yodprasit and Ngarmnil in [85]. The active inductor consists of a differentially configured transconductor with a positive transconductance and a common-source transconductor with a negative transconductance.

A. Lossless Karsilayan-Schaumann Active Inductors

The inductance of the inductor can be derived from the small-signal analysis of the inductor with the assumption $g_{ds} = 0$. The admittance of the active inductor is given by

$$Y = sC_{gs1}\frac{sC_{gs2} + g_{m2}}{s(C_{gs1} + C_{gs2}) + (g_{m1} + g_{m2})}$$

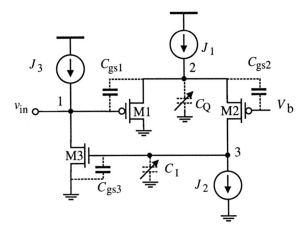

Figure 2.40.   Simplified schematic of Karsilayan-Schaumann active inductor.

$$+ \frac{g_{m2}g_{m3}}{sC_{gs3}} \frac{sC_{gs1} + g_{m1}}{s(C_{gs1} + C_{gs2}) + (g_{m1} + g_{m2})}. \qquad (2.92)$$

Further assume M_1 and M_2 are perfectly matched, i.e. $g_{m1} = g_{m2} = g_m$ and $C_{gs1} = C_{gs2} = C_{gs}$. Eq.(2.92) is simplified to

$$Y = \frac{1}{s\left(\dfrac{2C_{gs3}}{g_m g_{m3}}\right)} + s\left(\frac{C_{gs}}{2}\right). \qquad (2.93)$$

Eq.(2.93) reveals that the active inductor can be represented by a capacitor in parallel with an inductor. The capacitance and inductance are given by

$$C_p = \frac{C_{gs}}{2},$$
$$\qquad\qquad (2.94)$$
$$L = \frac{2C_{gs3}}{g_m g_{m3}}.$$

It is interesting to note that the preceding results can also be obtained using the results given in (2.9) directly. The differentially-configured transconductor with only one of its two input terminals is connected to the input has a transconductance $\frac{g_m}{2}$. The capacitance encountered at the input node of the active inductor is given by $\frac{C_{gs}}{2}$ as capacitors of C_{gs1} and C_{gs2} are connected in series.

Eq.(2.94) reveals that the inductance of the active inductor can be increased by increasing the capacitance between the gate and source of M_3. This can be achieved by adding an auxiliary capacitor C_I in parallel with C_{gs3}, as shown in Fig.2.40. The inductance in this case becomes

$$L = \frac{2(C_{gs3} + C_I)}{g_m g_{m3}}. \tag{2.95}$$

The auxiliary capacitor C_I can be implemented using MOS varactors. The inductance of the active inductor can be tuned in this way.

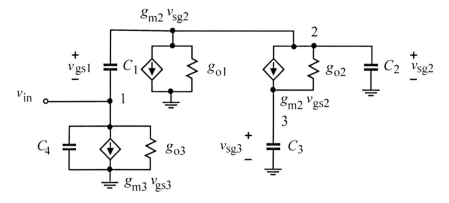

Figure 2.41.   Small-signal equivalent circuit of Karsilayan-Schaumann active inductor.

B. Lossy Karsilayan-Schaumann Active Inductors

The preceding analysis excludes the effect of g_{ls} and other parasitic capacitances of the transistors. Due to the absence of the lossy conductance g_{ls}, the quality factor of the active inductor can not be studied. In what follows, we investigate the quality factor of the active inductor and its tunability by following the approach of Karsilayan and Schaumann. The small-signal equivalent circuit of the active inductor is shown in Fig.2.41 where C_1, C_2, C_3, and C_4 represent the total capacitances including both intrinsic and parasitic capacitances encountered at or between respective nodes. It was shown in [74] that the admittance of the active inductor is given by

$$Y \approx \frac{1}{j\omega\left(\dfrac{2C_3}{g_m g_{m3}}\right) + \dfrac{G(\omega)}{g_m g_{m3}}}, \tag{2.96}$$

where

$$G(\omega) = g_{o2} + 2g_{o4} - \omega^2 C_3\left(\frac{C_1 + C_2}{g_m}\right). \tag{2.97}$$

The active inductor can be represented by an inductor of inductance

$$L = \frac{2C_3}{g_m g_{m3}}$$

(2.98)

in series with a resistor of resistance

$$R_s = \frac{G(\omega)}{g_m g_{m3}}.$$

(2.99)

The quality factor of the active inductor is obtained from

$$Q \approx \frac{\omega L}{R_s} = \frac{2\omega C_3}{G(\omega)}.$$

(2.100)

It is observed from (2.100) that if we set $G(\omega) = 0$, i.e.

$$C_2 = \frac{(g_{o2} + 2g_{o4})g_m}{\omega^2 C_3} - C_1,$$

(2.101)

the quality factor of the active inductor will become infinite. To achieve this, an auxiliary capacitor C_Q can be added at the source of M_1 and M_2, as shown in Fig.2.40. An important observation is that Q is tuned by varying C_Q, which is the capacitance encountered at the source of M_1 and M_3 while the inductance of the active inductor is tuned by varying C_I, the capacitance of the auxiliary capacitor added between the gate and source of M_3. In other words, Q and L can be tuned independently.

It was demonstrated in [74] that the quality factor of the active inductor was made nearly 400 and the inductance of the active inductor exceeded 600 nH in a 0.5μm CMOS implementation of Karsilayan-Schaumann active inductor.

C. Variations of Karsilayan-Schaumann Active Inductors

To increase the speed of the active inductor and to reduce the silicon consumption, it was shown by Xiao and Schaumann in [61] that the preceding Karsilayan-Schaumann active inductor can also be implemented using all nMOS transistors (excluding biasing current sources), as shown in Fig.2.42. Implemented in TSMC-0.18μm CMOS technology, this active inductor exhibited a self-resonant frequency of 6.68 GHz and a quality factor of 106.

In [86, 87], the common-source configured transconductor of the preceding Karsilayan-Schaumann active inductor was replaced with a static inverter to boost the transconductance of the transconductor from g_{n3} to $g_{m3} + g_{m4}$, as shown in Fig.2.43. The inductance is tuned by varying C_Q while the quality

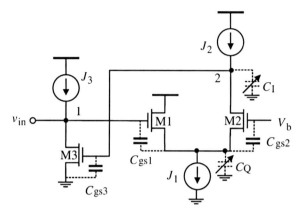

Figure 2.42. Simplified schematic of Karsilayan-Schaumann active inductor using nMOS transistors only.

factor is tuned by changing C_I. The input voltage of the static inverter must satisfy $V_{IL} \leq v_2 \leq V_{IH}$ where V_{IL} and V_{IH} are the lower and upper voltage bounds of the transition region of the static inverter, respectively, in order to ensure that M_3 and M_4 are in the saturation. A disadvantage of this design is the stringent constraint imposed on the voltage swing of node 2 of the active inductor.

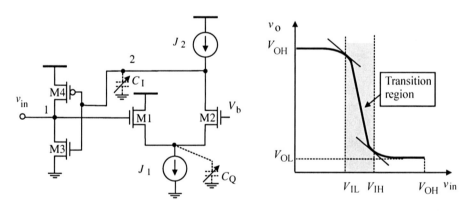

Figure 2.43. Simplified schematic of Karsilayan-Schaumann active inductor using a static-inverter negative transconductor.

2.3.8   Yodprasit-Ngarmnil Active Inductors

It was pointed out earlier that to boost the quality factor, the effect of both R_s and R_p of an active inductor must be compensated for. The $L \sim R_s$ series branch of the RLC equivalent circuit of an active inductor can be replaced with the $L \sim \hat{R}_p$ parallel branch shown in Fig.2.44(b) with the inductance and the shunt resistance given by

$$\hat{L} = L(1 + \frac{1}{Q^2}),$$

$$\hat{R}_p = R_p(1 + Q^2).$$

(2.102)

Both $\hat{R}_p$ and R_p can be combined into a single parallel resistor of resistance $R_{p,total} = \hat{R}_p \| R_p$.

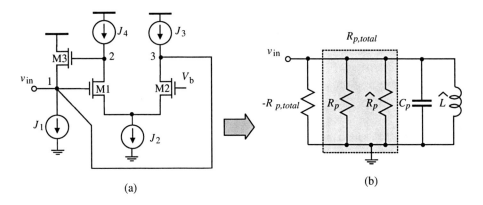

Figure 2.44. Simplified schematic of Yodprasit-Ngarmnil active inductor.

To boost the quality factor, a negative resistor of resistance $-R_{p,total}$ can be connected in parallel with $R_{p,total}$, as shown in Fig.2.44(b), such that the net resistive loss of the active inductor vanishes. It was also shown earlier that negative resistors are realized by using positive feedback. In Fig.2.44(a), the added electrical connection between the input terminal of the active inductor and the drain of M_2 forms the needed positive feedback. This is because an increase in v_{in} will result in an increase in i_{D1}, subsequently a decrease in i_{D2} as $i_{D1} + i_{D2} = J_2$ (constant). This is echoed with an increase in the drain voltage of M_2, which will further increase v_{in}. The impedance looking into the gate of M_1 at low frequencies is given by

$$Z_{in} \approx -\left(\frac{1}{g_{m1}} + \frac{1}{g_{m2}}\right).$$

(2.103)

The preceding analysis reveals that the differential pair offers two distinct functions simultaneously :

- It behaves as a transconductor with a negative transconductance to construct the gyrator-C active inductor.

- It provides the needed negative resistance between the input terminal and the ground to cancel out the parasitic resistances of the active inductor.

When g_o is considered, it was shown in [85] that the quality factor of Yodprasit-Ngarmnil active inductor is given by

$$Q = \frac{\sqrt{g_{m3}g_{m1}C_{gs3}C_{gs1}}}{\dfrac{C_{gs1}}{r_{o1}} + \dfrac{2C_{gs3}}{r_{o3}}}. \tag{2.104}$$

Eq.(2.104) shows that Q can be tuned by either changing $g_{m1,2}$ or $r_{o1,2}$. Because the former also changes the inductance of the inductor, the preferred choice is therefore to vary r_o of M_1 and M_2, which can be achieved by employing the cascode configuration of the differential-pair transconductor and varying the gate voltage of M_4 and M_5, as shown in Fig.2.45. r_{o1} and r_{o2} in Fig.2.45 now become $(g_{m4}r_{o4})r_{o1}$ and $(g_{m5}r_{o5})r_{o2}$, respectively. Transconductances $g_{m4,5}$ can be tuned by varying the gate voltage of $M_{4,5}$. As demonstrated in [85], Q was tuned up to 12000 in a 0.6μm implementation of the active inductor.

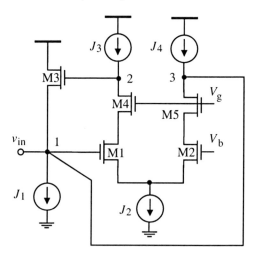

Figure 2.45. Simplified schematic of Cascode Yodprasit-Ngarmnil active inductors.

2.3.9   Uyanik-Tarim Active Inductor

The CMOS active inductor proposed by Uyanik and Tarim in [88] with its simplified schematic shown in Fig.2.46 only has two transistors connected in series between V_{DD} and ground rails, making it very attractive for low-voltage applications. As seen in Fig.2.46, M_1 and J form a transconductor with a negative transconductance g_{m1}. $M_{2,3,4}$ form a transconductor with a positive transconductance $\dfrac{g_{m2}g_{m4}}{g_{m3}} = g_{m2}$, provided that M_3 and M_4 are identical. To tune the inductance of the active inductor without affecting the parasitic series resistance of the active inductor, which controls the quality factor of the active inductor, a varactor C is added between the gate of M_2 and the ground.

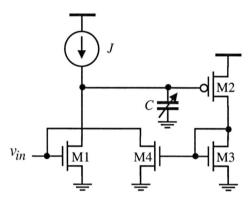

Figure 2.46. Simplified schematic of Uyanik-Tarim active inductor.

Implemented in UMC-0.13μm 1.2V CMOS technology, the simulation results in [88] shows that the active inductor had a wide frequency range from 0.3 GHz to its self-resonant frequency of approximately 7.32 GHz. The quality factor of the active inductor exceeded 100 in the frequency range 4.8-6.4 GHz with its phase error less than 1 degree. The maximum quality factor was 3900, occurring at 5.75 GHz. The minimum number of transistors stacked between the power and ground rails also enabled the active inductor to have a large input signal swing of 18 mV. The inductance was from 38 nH to 144 nH.

2.3.10 Carreto-Castro Active Inductors

The BiCMOS active inductors proposed by Carreto-Castro in [89] can also be implemented in CMOS technologies, as shown in Fig.2.47. Neglecting C_{gd} and g_o of the transistor. The input impedance of nMOS Carreto-Castro active inductor is given by

$$Z = \frac{1}{g_m} \frac{sRC_{gs}+1}{s\dfrac{C_{gs}}{g_m}+1}. \tag{2.105}$$

The zero of $Z(s)$ is at frequency

$$\omega_z = \frac{1}{RC_{gs}} \tag{2.106}$$

and the pole of $Z(s)$ is at frequency

$$\omega_p = \frac{g_m}{C_{gs}}. \tag{2.107}$$

The inductance of Carreto-Castro active inductor can be derived by examining the input admittance

$$
\begin{aligned}
Y &= \frac{sC + g_m}{sRC + 1} \\
&= \frac{1}{R} + \frac{1}{s\left(\dfrac{RC}{g_m - \frac{1}{R}}\right) + \dfrac{1}{g_m - \frac{1}{R}}}.
\end{aligned} \tag{2.108}
$$

It is seen from (2.108) that the parameters of the RLC equivalent circuit of the active inductor are given by

$$R_p = R,$$

$$L = \frac{R^2 C}{Rg_m - 1}, \tag{2.109}$$

$$R_s = \frac{R}{Rg_m - 1}.$$

In order to have $L > 0$ and $R_s > 0$,

$$Rg_m > 1 \tag{2.110}$$

is required. This condition also ensures that $\omega_z < \omega_p$. In the frequency range $\omega_z < \omega < \omega_p$, the circuit is inductive. Because

$$\frac{\omega_p}{\omega_z} = Rg_m, \tag{2.111}$$

for practical applications, $Rg_m \gg 1$ is usually required to maximize the effective frequency range. In this case

$$L \approx \frac{RC}{g_m} = \frac{R}{\omega_t},$$

$$(2.112)$$

$$R_s \approx \frac{1}{g_m}.$$

It is interesting to note that Hara active inductors, Wu folded active inductors, and Carreto-Castro active inductors all have the same expressions for L and R_s.

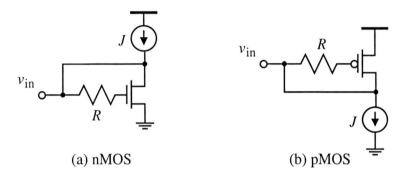

(a) nMOS   (b) pMOS

Figure 2.47. Simplified schematic of Carreto-Castro active inductors.

Figs.2.48 and 2.49 show the dependence of the resistance and reactance of a nMOS Carreto-Castro active inductor on R and J. The active inductor was implemented in TSMC-0.18μm 1.8V CMOS technology and analyzed using SpectreRF with BSIM3V3 device models.

2.3.11   Thanachayanont-Payne Cascode Active Inductors

It was pointed out earlier that to maximize the frequency range of active inductors, ω_z should be minimized and ω_o should be maximized. Maximizing ω_o is rather difficult because ω_o of active inductors is set by the cut-off frequency of the transconductors constituting the active inductors. The frequency of the zero of active inductors given by $\omega_z = \dfrac{g_{o1}}{C_{gs2}}$, on the other hand, can be lowered by either increasing C_{gs2} or decreasing g_{o1}. The former is at the cost of lowering ω_o and should therefore be avoided. To reduce g_{o1}, Thanachayanont and Payne proposed the cascode active inductors shown in Fig.2.50 [90, 65, 73]. Cascode can be implemented in either of the two transconductors, as shown in Fig.2.50. Note that a modification in the polarity of the transconductors is required to ensure the existence of a negative feedback in the gyrator-C configuration.

The impedance of Thanachayanont-Payne cascode active inductor is given by

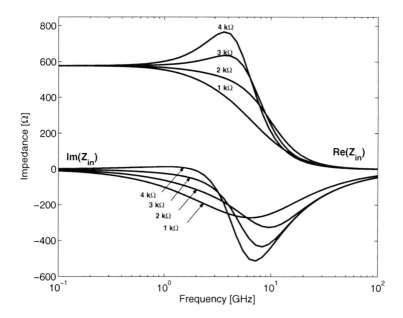

Figure 2.48. Dependence of Z of nMOS Carreto-Castro active inductor on R. Circuit parameters : $L = 0.18\mu$m, $W = 10\mu$m, $R = 2k\Omega$.

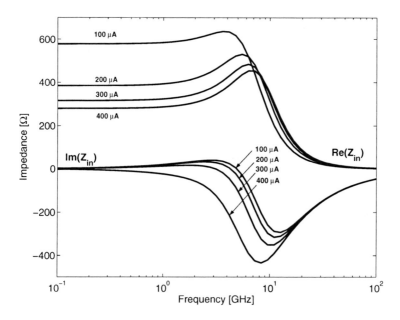

Figure 2.49. Dependence of Z of nMOS Carreto-Castro active inductor on J. Circuit parameters : $L = 0.18\mu$m, $W = 10\mu$m, $R = 2k\Omega$.

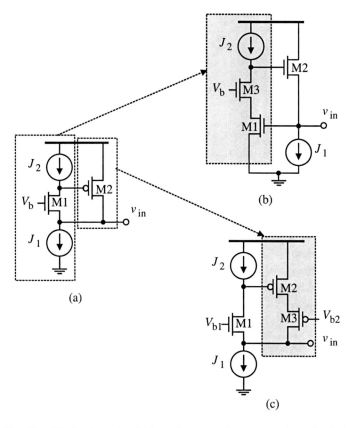

Figure 2.50. Simplified schematic of Thanachayanont-Payne cascode active inductors. (a) Basic active inductor. (b,c) Thanachayanont-Payne cascode active inductors.

$$Z \approx \left(\frac{g_{o1}g_{o3}}{C_{gs1}C_{gs2}g_{m3}} \right) \frac{s\left(\dfrac{C_{gs2}g_{m3}}{g_{o1}g_{o3}} \right) + 1}{s^2 + s\left(\dfrac{g_{o1}g_{o3}}{C_{gs2}g_{m3}} + \dfrac{g_{o1}}{C_{gs1}} \right) + \dfrac{g_{m1}g_{m2}}{C_{gs1}C_{gs2}}}, \qquad (2.113)$$

where $g_m \gg g_o$ was utilized in simplifying the results. It is seen from (2.113) that the frequency of the zero of Z is reduced from

$$\omega_z = \frac{g_{o1}}{C_{gs2}} \qquad (2.114)$$

without the cascode to

$$\omega_z = \left(\frac{g_{o1}}{C_{gs2}}\right)\frac{1}{g_{m3}r_{o3}} \tag{2.115}$$

with the cascode. The self-resonant frequency of the active inductor given by

$$\omega_o = \sqrt{\frac{g_{m1}g_{m2}}{C_{gs1}C_{gs2}}} = \sqrt{\omega_{t1}\omega_{t2}}, \tag{2.116}$$

however, remains unchanged. It should not be surprised to see that cascodes do not change ω_o. This is because cascodes are not subject to Miller effect and has no effect on the bandwidth. Cascode configurations thus can effectively expend the frequency range of active inductors by lowering the lower bound of the frequency range of active inductors.

The parameters of the RLC equivalent network of the cascode active inductor can be obtained by examining the admittance of the cascode inductor and the results are given by

$$R_p = \frac{1}{g_{o2}},$$

$$C_p = C_{gs1},$$

$$R_s = \left(\frac{g_{o1}}{g_{m1}g_{m2}}\right)\frac{1}{g_{m3}r_{o3}}, \tag{2.117}$$

$$L = \frac{C_{gs2}}{g_{m1}g_{m2}}.$$

It is evident from (2.117) that the cascode active inductor has the same inductance as that of the corresponding non-cascode gyrator-C active inductor. The parasitic series resistance is reduced from

$$R_s = \frac{g_{o1}}{g_{m1}g_{m2}} \tag{2.118}$$

without the cascode to

$$R_s = \frac{g_{o1}}{g_{m1}g_{m2}}\left(\frac{1}{g_{m3}r_{o3}}\right) \tag{2.119}$$

with the cascode. The parallel resistance is increased from

$$R_p = \frac{1}{g_{m1}} \tag{2.120}$$

without the cascode to

$$R_p = \frac{1}{g_{o2}}. \tag{2.121}$$

with the cascode. Both improve the quality factor of the active inductor, as is evident from

$$Q \approx \frac{\omega L}{R_s} = \omega C_{gs2} r_{o1} \quad (R_s \text{ dominates}) \tag{2.122}$$

or

$$Q \approx \frac{R_p}{\omega L} = \frac{g_{m2}}{\omega C_{gs2}} \quad (R_p \text{ dominates}) \tag{2.123}$$

without the cascode and

$$Q \approx \frac{\omega L}{R_s} = \omega C_{gs2} r_{o1} (g_{m3} r_{o3}) \quad (R_s \text{ dominates}) \tag{2.124}$$

or

$$Q \approx \frac{R_p}{\omega L} = (r_{o2} g_{m1}) \frac{g_{m2}}{\omega C_{gs2}} \quad (R_p \text{ dominates}) \tag{2.125}$$

with the cascode. To summarize, the cascode configurations of gyrator-C active inductors offer the following attractive characteristics :

- Frequency range expansion by lowering the lower bound of the frequency range.

- Quality factor improvement by lowering the parasitic series resistance R_s and increasing the parasitic parallel resistance R_p.

- No reduction in the upper bound of the frequency range.

- No reduction in the inductance.

- The minimum supply voltage of the active inductor without the cascode is given by $V_T + 2V_{sat}$. For the cascode active inductor of Fig.2.50(b), $V_{DD,min} = 2V_T + V_{sat}$. For the cascode active inductor of Fig.2.50(c), $V_{DD,min} = V_T + 2V_{sat}$.

2.3.12   Weng-Kuo Cascode Active Inductors

Weng and Kuo proposed a current-reuse cascode active inductor in [62] to eliminate the drawback of the preceding Thanachayanont-Payne cascode active inductors that the inductance and quality factor can not be tuned independently. The simplified schematic of Weng-Kuo active inductor is shown in Fig.2.51. It is seen that g_{m1} is proportional to $J_1 + J_3$ while g_{m3} is only proportional to J_1. Because $L = \dfrac{C_{gs2}}{g_{m1}g_{m2}}$, $R_s = \dfrac{g_{o1}g_{o3}}{g_{m1}g_{m2}g_{m3}}$, $C_p = C_{gs1}$, and $R_p = \dfrac{1}{g_{o2}}$, we have

$$\omega_o = \sqrt{\frac{g_{m1}g_{m2}}{C_{gs1}C_{gs2}}},$$

$$Q(\omega_o) = \frac{\omega_o L}{R_s} = \frac{g_{m3}}{g_{o1}g_{o3}}\sqrt{\frac{g_{m1}g_{m2}C_{gs2}}{C_{gs1}}}. \tag{2.126}$$

ω_o can be tuned by varying g_{m1} and g_{m2} while Q can be tuned by varying g_{m3} only. So the tuning of Q can be made independent of ω_o. Note that because the tuning of ω_o, however, will affect Q, an adjustment of Q is therefore required after each tuning of ω_o.

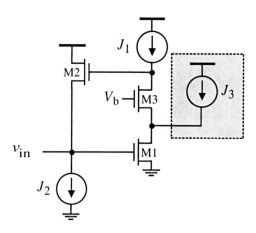

Figure 2.51.   Simplified schematic of Weng-Kuo active inductor.

2.3.13   Manetakis Regulated Cascode Active Inductors

The performance of the preceding cascode active inductors can be improved by further reducing R_s. This is achieved by using regulated cascodes and multi-regulated cascodes, as shown in Fig.2.52 [91]. The parameters of the

RLC equivalent circuit of the regulated cascode gyrator-C active inductor are given by

$$G_{o1} = \frac{1}{g_{o1}(r_{o3}g_{m3})(r_{o4}g_{m4})},$$

$$C_1 = C_{gs2},$$

$$G_{o2} = g_{o2},$$

$$C_2 = C_{gs1},$$

(2.127)

from which we obtain

$$R_p = \frac{1}{G_{o2}},$$

$$C_p = C_2,$$

$$R_s = \frac{G_{o1}}{g_{m1}g_{m2}},$$

$$L = \frac{C_1}{g_{m1}g_{m2}}.$$

(2.128)

Because $\omega_z = \dfrac{G_{o1}}{C_1}$ and $\omega_o = \dfrac{1}{\sqrt{LC_p}}$, the lower bound of the frequency is reduced while the upper bound of the frequency range remains unchanged.

The parameters of the RLC equivalent circuit of the multi-regulated cascode gyrator-C active inductor are given by

$$G_{o1} = \frac{1}{g_{o1}(r_{o3}g_{m3})(r_{o4}g_{m4})(r_{o5}g_{m5})},$$

$$C_1 = C_{gs2},$$

$$G_{o2} = g_{o2},$$

$$C_2 = C_{gs1}.$$

(2.129)

It is evident from (2.129) that the lower bound of the frequency is further reduced while the upper bound of the frequency range remains unchanged.

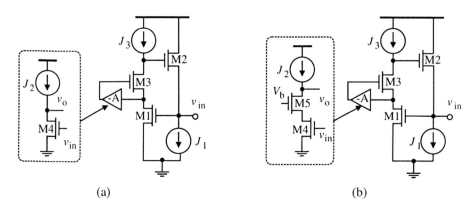

Figure 2.52. Simplified schematic of Manetakis regulated and multi-regulated cascode active inductors. (a) Regulated cascode active inductors. (b) Multi-regulated cascode active inductors.

The simulation results of cascode active inductors, regulated cascode active inductors and multi-regulated cascode active inductors given in [91] demonstrated that regulated cascode reduced the lower frequency bound of active inductors by one decade while multi-regulated cascode further reduced the lower frequency bond by more than one decade.

2.3.14 Hsiao Feedback Resistance Cascode Active Inductors

It was shown earlier that cascode active inductors offer a large frequency range and a high quality factor. To further improve the quality factor, feedback resistance active inductors shown in Fig.2.53 were proposed by Hsiao *et al.* in [92]. This type of active inductors was further investigated in [93, 94]. By assuming that the biasing current source transistors M_n and M_p are ideal, the parameters of the RLC equivalent circuit of the active inductor are given by

$$C_p = C_{gs3},$$

$$R_p = \frac{R_f g_{o2} + 1}{2g_{o2} + R_f g_{o2}^2},$$

$$R_s = \frac{g_{m1}g_{o2}g_{o3} + \omega^2[g_{m2}C_{gs1}^2 - g_{m1}C_{gs1}C_{gs2}(R_f g_{o2} + 1)]}{g_{m1}^2 g_{m2}g_{m3} + \omega^2 g_{m2}g_{m3}C_{gs1}^2}, \qquad (2.130)$$

$$L = \frac{g_{m1}g_{m2}C_{gs1} + \omega^2 C_{gs1}^2 C_{gs2}(R_f g_{o2} + 1)}{g_{m1}^2 g_{m2}g_{m3} + \omega^2 g_{m2}g_{m3}C_{gs1}^2}.$$

It is seen from (2.130) that the added feedback resistor R_f lowers R_s and increases L simultaneously. Both boost the quality factor of the active inductor. The resistance of the feedback resistor R_f can be tuned by connecting a nMOS transistor in parallel with a poly resistor, as shown in the figure. Note R_f does not consume any static power. For the special case where $R_f = 0$, Eq.(2.130) becomes

$$C_p = C_{gs3},$$

$$R_p \approx \frac{1}{g_{o2}},$$

$$R_s \approx \frac{g_{o2}g_{o3}}{g_{m1}g_{m2}g_{m3}}, \tag{2.131}$$

$$L = \frac{C_{gs2}}{g_{m1}g_{m2}}.$$

It should be noted that the self-resonant frequency of the active inductor with the feedback resistors R_f is reduced due to the increase of L. It was shown in [94] that the decrease of the self-resonant frequency of the feedback resistance active inductors can be compensated for by varying the biasing voltage V_b of the current-source transistor M_p. The simulation results of a feedback resistance active inductor implemented in a 0.18μm CMOS technology showed that the inductance of the inductor was 15 nH with the quality factor exceeding 50 and the self-resonant frequency of several GHz [92–94].

The preceding Hsiao feedback resistance cascode active inductors were further developed by Liang *et al.* where the cascode portion of the active inductor is replaced with a regulated cascode branch, as shown in Fig.2.54 so that the advantages of the regulated cascodes active inductor investigated earlier can be utilized [50].

2.3.15   Abdalla Feedback Resistance Active Inductors

Karsilayan-Schaumann active inductors studied earlier offer the key advantage of the independent tuning of their inductance and quality factor. In [95], Abdalla *et al.* modified Karsilayan-Schaumann active inductors by adding a feedback resistor between the two transconductors of the active inductor to improve the quality factor of the inductor, as shown in Fig.2.55. The added feedback resistor increases the inductance of the active inductor and at the same time lowers the parasitic series resistance of the active inductor, thereby boosting the quality factor of the active inductor. C_I and C_Q are MOS varactors to tune the inductance and quality factor of the active inductor, respectively.

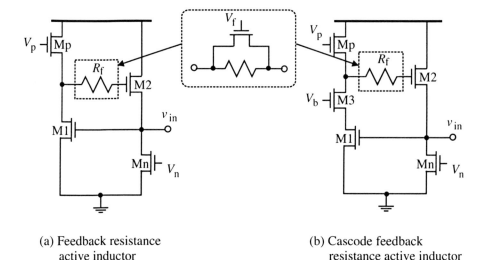

(a) Feedback resistance
active inductor

(b) Cascode feedback
resistance active inductor

Figure 2.53.   Simplified schematic of Hsiao feedback resistance active inductors.

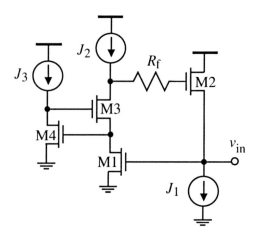

Figure 2.54.   Simplified schematic of Liang feedback resistance regular cascode active inductors.

2.3.16   Nair Active Inductors

Wei *et al.* proposed a MESFET-version high-Q active inductor with loss compensated by a feedback network in [96]. This active inductor was modified and implemented in CMOS by Nair *et al.* in [97] and was used in design of a low-power low-noise amplifier for ultra wideband applications. The simplified schematic of Nair active inductor is shown in Fig.2.56. It is a cascode active inductor with a negative feedback network consisting of R_1, R_2, and C_2. It was shown in [97] that the parameters of this active inductor are given by

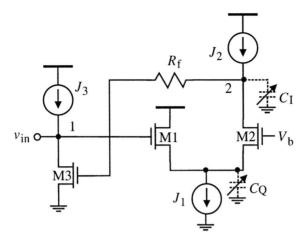

Figure 2.55. Simplified schematic of single-ended Abdalla feedback resistance active inductors.

$$L = \frac{C_3 + C_{gs3}}{g_{m1}g_{m3}},$$

$$\omega_o = \sqrt{\frac{g_{m1}g_{m3}}{C_{gs1}(C_3 + C_{gs3})}}, \qquad (2.132)$$

$$Q(\omega_o) = \sqrt{\frac{g_{m1}g_{m3}(C_3 + C_{gs3})}{C_{gs1}g_{o1}^2}}.$$

It is seen from (2.132) that C_3 is used to boost the inductance. It also increases the quality factor and decreases the self-resonant frequency of the active inductor. The RC network consisting of $R_1, R_2 - C_2$ is a negative feedback network that reduces the parasitic resistances of the active inductor.

2.3.17 Active Inductors with Low Supply-Voltage Sensitivity

It was pointed out earlier that the parameters of gyrator-C active inductors, such as the inductance and parasitic resistances, are sensitive to the fluctuation of the supply voltage of the active inductors. This is because a V_{DD} fluctuation not only alters the dc operating point of the transconductors constituting the active inductors, it also changes the junction capacitances of the active inductors. Replica-biasing is an effective means to reduce the effect of supply voltage fluctuation on the inductance of active inductors. Fig.2.57 shows the configuration of Wu current-reuse active inductors with replica biasing [66]. The

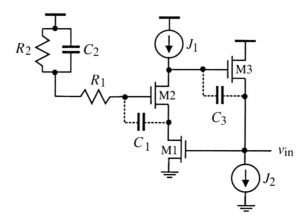

Figure 2.56.   Simplified schematic of Nair active inductors.

replica-biasing section consists of a sensing circuit made of $M_{4,5,6}$ and an auxiliary voltage amplifier. An increase in V_{DD} will lead to an increase in $v_{SG3,6}$, subsequently the channel current of $M_{3,6}$. The voltage of the non-inverting terminal of the amplifier will also increase. The output of the amplifier will increase the gate voltage of $M_{3,6}$, which ensures that $v_{SG3,6}$ is kept unchanged approximately, minimizing the effect of V_{DD} fluctuation. The width of the transistors in the replica-biasing section should be the same as that of the active inductor section so that both will sense the same voltage change caused by the variation of V_{DD}.

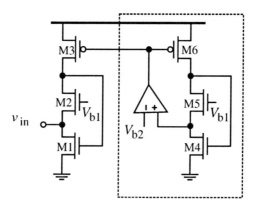

Figure 2.57.   Simplified schematic of Wu current reuse active inductor (nMOS) with replica biasing.

2.4 Implementation of Differential Active Inductors

2.4.1 Lu Floating Active Inductors

The schematic of the active inductor proposed by Lu *et al.* is shown in Fig.2.58 [60]. It is a differentially configured gyrator-C active inductor. Transistors M_5 and M_6 are biased in the triode region and behave as voltage-controlled resistors whose resistances are controlled by the gate voltage V_b. All other transistors are biased in the saturation.

The negative feedback of the active inductor is as the followings : an increase in v_{1+} and a decrease in v_{1-} will result in an increase in v_{2+} and a decrease in v_{2-} due to the common-gate operation of $M_{1,2}$. The source followers of $M_{3,4}$ ensure that v_{1+} and v_{1-} will be reduced accordingly by approximately the same amount.

To find out the parameters of the RLC equivalent circuit of the active inductor, we represent M_5 and M_6 with channel conductances g_{ds5} and g_{ds6}, respectively, in the small-signal equivalent circuit of the active inductor. It can be shown that the differential input impedance of the inductor is given by

$$Z = \frac{2[s(C_{gs1} + C_{gs3}) - g_{m1} + g_{ds5}]}{g_{ds5}[g_{m1} + g_{m3} + s(C_{gs1} + C_{gs3})]}. \tag{2.133}$$

The parameters of the RLC equivalent circuit of the active inductor are given by

$$R_p = \frac{2}{g_{ds5}},$$

$$R_s = \frac{2(g_{ds5} - g_{m1})}{g_{ds5}(2g_{m1} + g_{m3} - g_{ds5})}, \tag{2.134}$$

$$L = \frac{2(C_{gs1} + C_{gs3})}{g_{ds5}(2g_{m1} + g_{m3} - g_{ds5})}.$$

Note that $2g_{m1} + g_{m3} > g_{ds5}$ and $g_{ds5} > g_{m1}$ are required in order for R_s and L to have a positive value. The preceding development reveals that the inductance of the active inductor can be tuned by varying V_b, subsequently g_{ds5}. An increase in V_b will push M_5 and M_6 from the triode region towards the saturation region, lowering g_{ds5} and g_{ds6}. This is echoed with an increase in the inductance of the inductor.

2.4.2 Grözing Floating Active Inductors

As pointed out earlier that floating gyrator-C active inductors can be constructed using a pair of differential transconductors. One of the simplest differ-

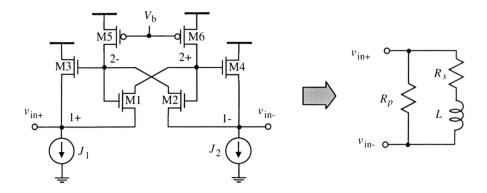

Figure 2.58.   Simplified schematic of Lu floating active inductor.

entially configured transconductors is the basic differential pair. The floating
active inductor proposed in [43, 44] and shown in Fig.2.59 employs two basic
differential-pair transconductors. Two negative resistors are connected across
the output nodes of the transconductors to cancel out the parasitic resistances
of the active inductor. The inductance of the active inductor is tuned by varying
the transconductances of the transconductors. This is done by adjusting the tail
currents of the differential pairs $J_{1,2}$. The quality factor of the active inductor
is tuned by varying the resistances of the negative resistors. This is attained by
changing the tail current sources $J_{3,4}$ of the negative resistors. It was shown
in [43, 44] that in a 0.18μm implementation of the active inductor, the quality
factor of the active inductor was 600 at 2 GHz while the self-resonant frequency
of the active inductor was 5.6 GHz

2.4.3   Thanachayanont Floating Active Inductors

The floating active inductor proposed in [37, 38, 98, 99, 65, 73] is shown
in Fig.2.60. It consists of two cascode-configured gyrator-C active inductors
investigated earlier. Transistors $M_{4,5}$ form a negative resistor to cancel out the
parasitic resistances of the active inductor so as to boost its quality factor. Note
that the resistance of the negative resistor can not be tuned in this implementation. The inductance of the active inductor is tuned by varying J_2. The active
inductor implemented in a 0.35μm CMOS technology offered an inductance of
70 nH and a self-resonant frequency of 2.8 GHz. The quality factor exceeded
100 over the frequency range from 0.83 GHz to 1.33 GHz with its maxima of
1970 at 1.08 GHz.

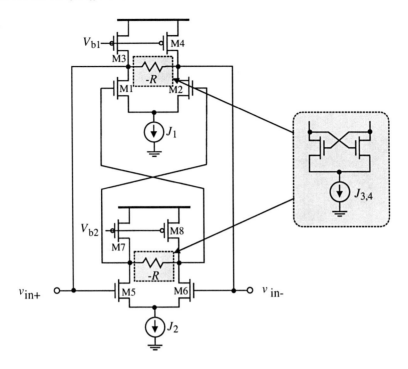

Figure 2.59. Simplified schematic of Grözing floating active inductor.

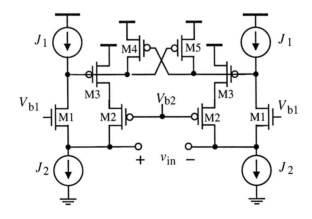

Figure 2.60. Simplified schematic of Thanachayanont floating active inductor.

2.4.4 Mahmoudi-Salama Floating Active Inductors

The floating active inductor proposed by Mahmoudi and Salama was used in the design of quadrature down converter for wireless applications [56, 100]. The schematic of Mahmoudi-Salama floating active inductor is shown in Fig.2.61. It consists of a pair of differential transconductors and a pair of negative resistors at the output of the transconductors. $M_{8,16}$ are biased in the triode and behave as voltage-controlled resistors. They are added to the conventional cross-coupled configuration of negative resistors to provide the tunability of the resistance of the negative resistors without using a tail current source.

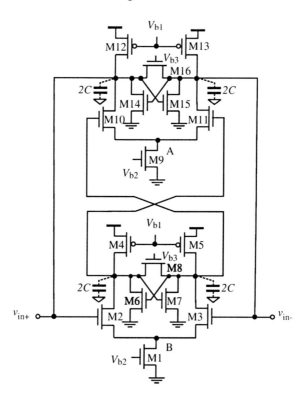

Figure 2.61.   Simplified schematic of Mahmoudi-Salama floating active inductor.

The small-signal equivalent circuit of the tunable negative resistor is shown in Fig.2.62 where a test voltage source V_x is added for the derivation of the equivalent resistance of the negative resistor. R represents the resistance of M_8. Writing KCL at nodes 1 and 2 yields

$$g_{m1}V_2 + \frac{V_x}{R} - I_x = 0 \qquad \text{(node 1)},$$

$$g_{m1}(V_x + V_2) - \frac{V_x}{R} + I_x = 0 \quad \text{(node 2)}. \tag{2.135}$$

The resistance of the negative resistors at low frequencies is obtained from (2.135)

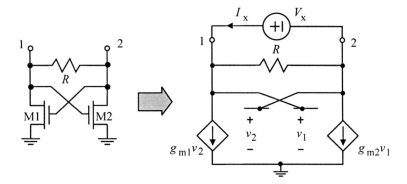

ht

Figure 2.62. Small-signal equivalent circuit of Mahmoudi-Salama floating active inductor at low frequencies. g_o of the transistors is neglected.

$$Z = \frac{V_x}{I_x} = -\frac{R\left(\dfrac{1}{g_{m1}} + \dfrac{1}{g_{m2}}\right)}{R - \left(\dfrac{1}{g_{m1}} + \dfrac{1}{g_{m2}}\right)}$$

$$= R // \left[-\left(\frac{1}{g_{m1}} + \frac{1}{g_{m2}}\right)\right]. \qquad (2.136)$$

The inductance of the active inductor is given by $L = \dfrac{C}{G_{m1}G_{m2}}$, where $2C$ is the total capacitance encountered at each of the output nodes of the transconductor, G_{m1} and G_{m2} are the transconductances of the differential transconductors 1 and 2, respectively. By assuming that nodes A and B are the virtual ground and neglecting C_{gd} and the diffusion junction capacitances, we have $C \approx \dfrac{C_{gs2,3,10,11}}{2}$ and $G_m = g_{m2,3,10,11}$.

2.4.5 Feedback Resistance Floating Active Inductors

The feedback resistance technique studied earlier was also employed in the design of floating active inductors by Akbari-Dilmaghani *et al.* in [101] to improve the performance of these inductors. A similar approach was used by Abdalla *et al.* in design of high-frequency phase shifters [102]. This section investigates these active inductors.

The schematic of the feedback resistance floating active inductor proposed by Akbari-Dilmaghani *et al.* is shown in Fig.2.63. It consists of two basic differential-pair transconductors and two feedback resistors. The functionality of the added feedback resistors is the same as that of the single-ended active inductors investigated earlier, i.e. lowering the parasitic series resistance and increasing the inductance.

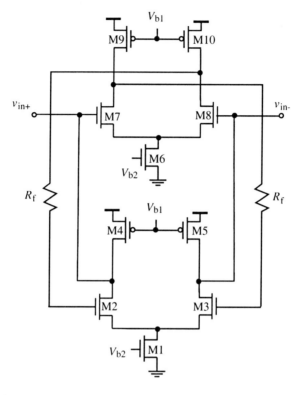

Figure 2.63. Simplified schematic of Akbari-Dilmaghani feedback resistance floating active inductor.

The simplified schematic of Abdalla differential feedback resistance floating active inductors is shown in Fig.2.64, $M_{11,12}$ are biased in the triode and behave as voltage-controlled resistors. It was shown in [102] that the inductance and the parasitic series resistance of the floating active inductor are given by

$$L = \frac{C + C_{gs4,5}\left(1 + \dfrac{R_f}{R_T}\right)}{g_{m1,2}g_{m4,5}},$$

$$R_s = \frac{\dfrac{1}{R_T} - \omega^2 C_{gs4,5}CR_f}{g_{m1,2}g_{m4,5}},$$

(2.137)

where

$$C = C_{gd7,8} + C_{db1,2} + C_{db7,8} + C_{gs1,2},$$

$$(2.138)$$

$$R_T = R_f \| R_{ds11,12} \| r_{o1,2} \| r_{o7,8}.$$

It is evident from (2.137) that R_f boosts L and lowers R_s simultaneously. Both improve the performance of the floating active inductor. Also seen from (2.137) and (2.138) that $M_{11,12}$ control the series resistance R_s of the active inductor. By adjusting V_{b1}, R_s can be minimized.

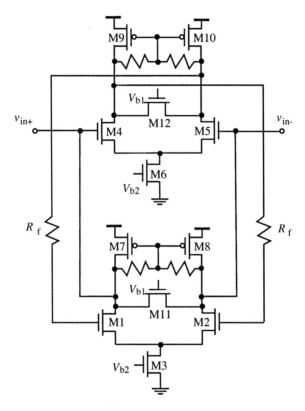

Figure 2.64. Simplified schematic differential Abdalla feedback resistance floating active inductors.

2.5 Class AB Active Inductors

The active inductors investigated up to this point fall into the category of class A active inductors. These class A active inductors suffer from a common drawback of a small voltage swing at the terminals of the active inductors, mainly due to the constraint that the input transistors of the transconductors of the active inductors should be biased and operated in the saturation. Note that although the input transistors of the transconductors of active inductors can be pushed into the triode region while still exhibiting an inductive characteristic, the nonlinearity of the active inductors will increase.

To increase the voltage swing of the active inductors without sacrificing linearity, class AB configurations can be employed. A class AB active inductor can be constructed from a nMOS-configured class A active inductor and a pMOS-configured class A active inductor, as shown in Fig.2.65 [65]. When the input voltage is high, only the nMOS class A active inductor is activated. When the input voltage is low, only the pMOS class A active inductor is activated, i.e. the two active inductors are operated in an interleave manner, depending upon the swing of the input voltage. The network thus exhibits an inductive characteristic over a large input voltage range. The swing of the input voltage and the minimum supply voltage of the class A and class AB active inductors are compared in Table 2.4. It is seen that the need for a large minimum supply voltage of class AB active inductors makes them less attractive for low-voltage applications. It should also be noted that when $V_T + V_{sat} \leq v_{in} \leq V_{DD} - (V_T + V_{sat})$, both the nMOS and pMOS class A active inductors are activated.

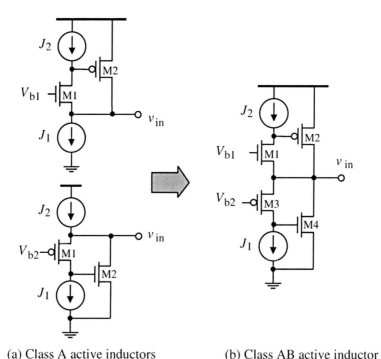

(a) Class A active inductors   (b) Class AB active inductor

Figure 2.65. Simplified schematic of Thanachayanont-Ngow class AB active inductor.

Table 2.4. A comparison of the input voltage swing and the minimum supply voltage of class A and class AB active inductors.

	Class A (nMOS)	Class A (pMOS)	Class AB
$V_{in,min}$	V_{sat}	$V_T + V_{sat}$	0
$V_{in,max}$	$V_{DD} - (V_T + V_{sat})$	$V_{DD} - V_{sat}$	V_{DD}
$V_{DD,min}$	$V_T + 2V_{sat}$	$V_T + 2V_{sat}$	$2V_T + 2V_{sat}$

Class AB active inductors can also be configured in cascodes to increase the frequency range and to boost the quality factor, as shown in Fig.2.66. Floating cascode class AB active inductors can also be constructed in a similar manner as that of non-cascode floating active inductors, as shown in Fig.2.67. Negative resistance compensation techniques can be employed to improve the performance of class AB active inductors.

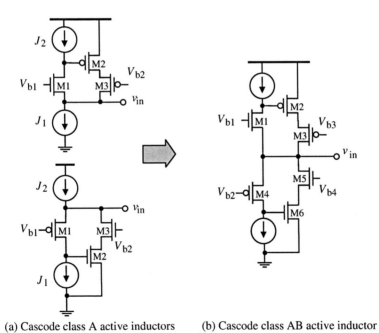

(a) Cascode class A active inductors   (b) Cascode class AB active inductor

Figure 2.66. Simplified schematic of Thanachayanont-Ngow cascode class AB active inductor.

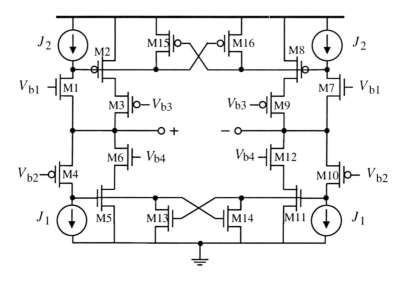

Figure 2.67. Simplified schematic of Thanachayanont-Ngow cascode floating class AB active inductors.

2.6   Chapter Summary

An in-depth examination of the principles, topologies, characteristics, and implementation of gyrator-C active inductors in CMOS technologies has been presented. We have shown that both single-ended and floating (differential) active inductors can be synthesized using gyrator-C networks. Lossless gyrator-C networks yield lossless active inductors while lossy gyrator-C networks synthesize lossy active inductors. To provide a quantitative measure of the performance of active inductors, a number of figure-of-merits have been introduced. These figure-of-merits include frequency range, inductance tunability, quality factor, noise, linearity, stability, supply voltage sensitivity, parameter sensitivity, signal sensitivity, and power consumption. Frequency range specifies the lower and upper bounds of the frequency in which gyrator-C networks are inductive. We have shown that the low frequency bound is set by the frequency of the zero of the gyrator-C networks while the upper frequency bound is set by the frequency of the pole of the networks. Both are due to the finite input and output impedances of the transconductors constituting active inductors. One of the key advantages of active inductors over their spiral counterparts is the large tunability of their inductance. We have shown that the inductance of gyrator-C active inductors can be tuned by varying either the transconductances of the transconductors or the load capacitance. The former is often used for the coarse control of the inductance whereas the latter is used for the fine control of the inductance. The distinct sensitivities of the quality factor of active inductors to their parasitic series and parallel resistances have been investigated. We have shown that the parasitic series resistance of an active inductor can be converted to an equivalent parallel parasitic resistance such that a negative shunt resistor can be employed to cancel out the effect of both the series and parallel parasitic resistances of the active inductor simultaneously. An emphasis has been given to the noise of active inductors as these inductors exhibit a high level of noise power. The input-referred noise generators of basic transconductors have been derived using the approach for noise analysis of 2-port networks whereas those of active inductors have been obtained using the approach for noise analysis of 1-port networks. The linearity of active inductors has been investigated. Because active inductors are active networks that are sensitive to both supply voltage fluctuation and parameter spread. We have used supply voltage sensitivity and parameter sensitivity to quantify the effect of these unwanted variations. A distinct characteristic of active inductors is the dependence of their performance on the swing of their signals. When active inductors are used in applications such as LC oscillators where signal swing is large, the effect of this dependence must be accounted for.

The second part of the chapter has focused upon the CMOS implementation of gyrator-C active inductors. The schematics and characteristics of single-ended and floating (differential) class A active inductors have been investigated in detail. Class AB gyrator-C active inductors have also been studied. The basic gyrator-C active inductors suffer from a small parasitic parallel resistance and a large parasitic series resistance due to the small output / input impedance of the transconductors constituting the active inductors. This in turn lowers the quality factor of the active inductors. To increase the parallel parasitic resistance and to lower the parasitic series resistance, transconductors with a large input / output resistance are critical. This can be achieved by changing the configuration of transconductors, such as Karsilayan-Schaumann active inductors that have a large output impedance, employing cascodes, or adding parallel negative resistors. To increase the tunability of active inductors, additional voltage-controlled capacitors can be employed at critical nodes of active inductors. This approach is often preferable over those that vary the transconductances of the transconductors of active inductors because the latter also alters other parameters of the active inductors such as inductance with a downside that the tuning range is rather small. Feedback resistance active inductors, both single-ended and floating, exhibit improved performance by increasing the inductance and decreasing the series parasitic resistance simultaneously. Both improve the quality of the active inductors. To increase the voltage swing of the active inductors without sacrificing linearity, class AB active inductors constructed from a pair of nMOS/pMOS-configured class A active inductors have been presented. Two main drawbacks of class AB active inductors are the circuit complexity of these inductors and their need for a high supply voltage.

Chapter 3

CMOS ACTIVE TRANSFORMERS

This chapter presents the principles, characteristics, and implementation of CMOS active transformers. Section 3.1 deals with the configurations of lossless gyrator-C active transformers where only the capacitor load of the transconductors synthesizing active transformers is considered. Both single-ended and floating lossless active transformers are investigated. The self and mutual inductances of the primary and secondary windings of lossless active transformers are investigated in detail. The configurations of gyrator-C active transformers with multiple windings are also developed. Lossy gyrator-C active transformers are developed with the consideration of both the resistive and capacitive loads of transconductors. The intrinsic relations between the self and mutual inductances are derived. Section 3.2 investigates the critical figure-or-merits that quantify the performance of active transformers. These figure-of-merits include stability, frequency range, inductance tunability, turn ratio, coupling factor, voltage and current transfer characteristics, impedance transformation, noise, quality factor, linearity, supply voltage sensitivity, parameter sensitivity, and power consumption. Section 3.3 looks into the circuit implementation of several CMOS active transformers and examines their characteristics. The chapter is summarized in Section 3.4.

3.1 Principles of Gyrator-C Active Transformers

Similar to spiral transformers that are constructed by coupling two spiral inductors via a magnetic link, a gyrator-C active transformer can be constructed by connecting two gyrator-C active inductors via a "magnetic" link, which is also synthesized using an electrical network. The two active inductors constituting the active transformer are referred to as the primary winding and the secondary winding of the transformer, depending upon the flow of the signals of the transformer. Gyrator-C active inductors are two-port networks and have two pairs of nodes, namely, the interface nodes and internal nodes. The coupling between the two gyrator-C active inductors of an active transformer can be established via either the interface nodes or the internal nodes of the active inductors.

3.1.1   Lossless Single-Ended Gyrator-C Active Transformers

A gyrator-C active transformer is said to be lossless if both the gyrator-C active inductors and the coupling network of the transformer are lossless, i.e. all transconductors have infinite input and output resistances. Fig.3.1 shows the configuration of lossless active transformers where the coupling between the primary and secondary winding active inductors is established from the interface nodes 1 and 2 to the internal nodes 3 and 4 of the active inductors. The synthesized transformer consists of

- A primary winding formed by the lossless gyrator-C active inductor $G_{m1} \sim C_1$.

- A secondary winding formed by the lossless gyrator-C active inductor $G_{m2} \sim C_2$.

- A coupling network that establishes an electrical connection between the primary and secondary winding active inductors. The coupling network should be such that there is no direct current path between the two active inductors such that the primary and secondary winding active inductors can be biased and tuned individually. There are four basic configurations that offer such a characteristic, namely voltage-controlled voltage sources, voltage-controlled current sources, current-controlled voltage sources, and current-controlled current sources. Voltage-controlled current sources, or transconductors are preferred due to their simple configurations and the fact that MOSFETs themselves are the simplest voltage-controlled current sources.

The coupling between the primary winding active inductor and the secondary winding active inductor of an active transformer can also be established from the internal nodes 3 and 4 to the interface nodes 1 and 2 of the two active inductors, as shown in Fig.3.2. An advantage of this approach is to relax the constraint imposed on the voltage swing of the coupling transconductors. In other words, the voltage swing of the primary and secondary windings of the transformer can be made large while that of the coupling transconductors remains small.

To derive the expression of the self and mutual inductances of the synthesized transformers, consider the lossless active transformer of Fig.3.2. Write KCL at nodes 1, 2, 3, and 4

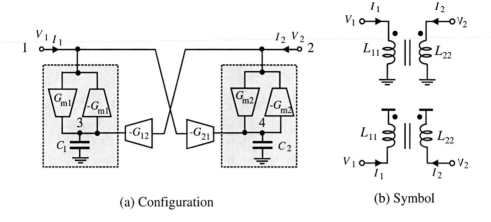

(a) Configuration   (b) Symbol

Figure 3.1. Lossless bi-directional gyrator-C active transformers. The active inductors are coupled from the interface nodes 1 and 2 to the internal nodes 3 and 4. Each active inductor is identified with a dotted box.

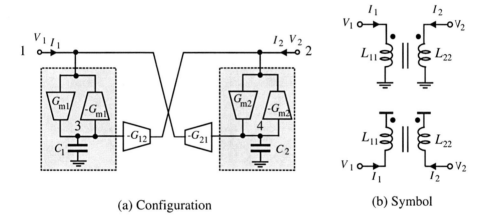

(a) Configuration   (b) Symbol

Figure 3.2. Lossless bi-directional gyrator-C active transformers. The active inductors are coupled from the internal nodes 3 and 4 to the interface nodes 1 and 2. Each active inductor is identified with a dotted box.

$$-I_1 + G_{m1}V_3 + G_{12}V_4 = 0 \quad \text{(node 1)},$$

$$-I_2 + G_{m2}V_4 + G_{21}V_3 = 0 \quad \text{(node 2)},$$

$$sC_1V_3 - G_{m1}V_1 = 0 \quad \text{(node 3)},$$

$$sC_2V_4 - G_{m2}V_2 = 0 \quad \text{(node 4)}.$$

$$(3.1)$$

Solve for V_2 from (3.1)

$$V_2 = \frac{sC_2}{G_{m2}^2 \Delta} I_2 + \left(\frac{G_{21}}{G_{m2}}\right) \frac{sC_1}{G_{m1}^2 \Delta} I_1, \tag{3.2}$$

where

$$\Delta = 1 - \frac{G_{12}G_{21}}{G_{m1}G_{m2}}. \tag{3.3}$$

The self inductance of the secondary winding, denoted by L_{22}, and the mutual inductance from the primary winding to the secondary winding, denoted by M_{21}, are obtained by comparing (3.2) with the voltage of the secondary winding of an ideal transformer given by

$$V_2 = sM_{21}I_1 + sL_{22}I_2 \tag{3.4}$$

and the results are

$$L_{22} = \frac{C_2}{G_{m2}^2 \Delta}, \tag{3.5}$$

$$M_{21} = \left(\frac{G_{21}}{G_{m2}}\right) \frac{C_1}{G_{m1}^2 \Delta}. \tag{3.6}$$

Similarly, the self inductance of the primary winding, denoted by L_{11}, and the mutual inductance from the secondary winding to the primary winding, denoted by M_{12}, are obtained by solving for V_1 from (3.1) and comparing the results with the voltage of the primary winding of an ideal transformer given by

$$V_1 = sL_{11}I_1 + sM_{12}I_2. \tag{3.7}$$

The results are

$$L_{11} = \frac{C_1}{G_{m1}^2 \Delta}, \tag{3.8}$$

and

$$M_{12} = \left(\frac{G_{12}}{G_{m1}}\right)\frac{C_2}{G_{m2}^2\Delta}. \tag{3.9}$$

Let us examine the results in detail :

- In order to have positive inductances, i.e. $L_{11}, L_{22}, M_{12}, M_{21} > 0, \Delta > 0$ or equivalently

$$G_{12}G_{21} < G_{m1}G_{m2} \tag{3.10}$$

 is required.

- If $G_{12}G_{21} \ll G_{m1}G_{m2}$, i.e. $\Delta \approx 1$, the self inductances L_{11} and L_{22} are independent of the transconductances G_{12} and G_{21} of the coupling transconductors. They can be tuned without affecting the coupling between the primary and secondary windings of the active transformer.

- If $G_{12}G_{21} \ll G_{m1}G_{m2}$, i.e. $\Delta \approx 1$, the mutual inductances M_{12} and M_{21} are linearly proportional to the transconductances G_{12} and G_{21} of the coupling transconductors. M_{12} and M_{21} can be tuned by only varying G_{12} and G_{21} without affecting the self inductances of the transformer.

- The relation between the mutual and self-inductances of the active transformer is given by

$$M_{12} = \frac{G_{12}}{G_{m1}}L_{22}, \tag{3.11}$$

and

$$M_{21} = \frac{G_{21}}{G_{m2}}L_{11}. \tag{3.12}$$

- The transconductors of the coupling network have a negative transconductance. If we change the sign of both G_{21} and G_{12} from negative to positive, the value of Δ will remain unchanged. As a result, L_{11} and L_{22} will remain unchanged. M_{12} and M_{21}, however, will change their signs accordingly. This observation reveals that active transformers with negative mutual inductances can be constructed by employing coupling transconductors with positive transconductances, as shown in Figs.3.3 and 3.4.

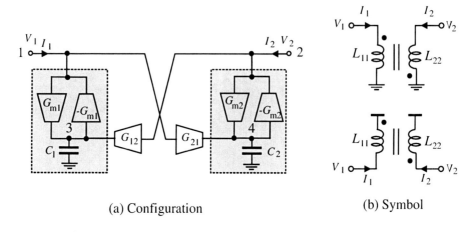

(a) Configuration   (b) Symbol

Figure 3.3. lossless bi-directional gyrator-C active transformers. The active inductors are coupled from the interface nodes 1 and 2 to the internal nodes 3 and 4. The transconductances of the coupling network are positive.

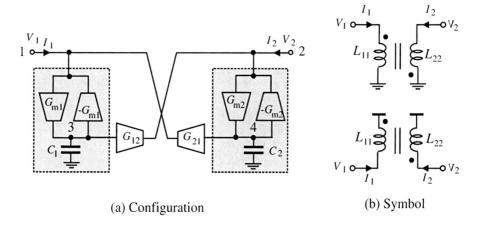

(a) Configuration   (b) Symbol

Figure 3.4. Lossless bi-directional gyrator-C active transformers. The active inductors are coupled from the internal nodes 3 and 4 to the interface nodes 1 and 2. The transconductances of the coupling network are positive.

3.1.2   Lossless Floating Gyrator-C Active Transformers

A lossless floating gyrator-C active transformer can be constructed using lossless floating (differential) active inductors and lossless floating (differential) coupling transconductors, as shown in Fig.3.5. Writing KCL at nodes 1, 2, 3, and 4 and simplifying the results yield

$$V_2^+ - V_2^- = \left(\frac{G_{12}}{G_{m2}}\right)\frac{C_1}{2G_{m1}^2\Delta}I_1 + \frac{sC_2}{2G_{m2}^2\Delta}I_2. \tag{3.13}$$

Eq.(3.13) reveals that the self inductance of the secondary winding and the mutual inductance from the primary winding to the secondary winding of the differential active transformer are given by

$$L_{22} = \frac{C_2}{2G_{m2}^2\Delta},$$

$$M_{21} = \left(\frac{G_{12}}{G_{m2}}\right)\frac{C_1}{2G_{m1}^2\Delta}. \tag{3.14}$$

In a similar manner, one can show that the self inductance of the primary winding and the mutual inductance from the secondary winding to the primary winding of the differential active transformer can be derived

$$L_{11} = \frac{C_1}{2G_{m1}^2\Delta},$$

$$M_{12} = \left(\frac{G_{21}}{G_{m1}}\right)\frac{C_2}{2G_{m2}^2\Delta}. \tag{3.15}$$

3.1.3 Lossy Single-Ended Gyrator-C Active Transformers

Active transformers become lossy when the active inductors constituting the active transformers are lossy, i.e. they have finite input or output resistances. In this section, we examine the characteristics of lossy gyrator-C active transformers where both the input and output capacitances and resistances of the transconductors synthesizing the active transformers are considered.

The block diagram of lossy gyrator-C transformers is shown in Fig.3.6. Note that the capacitances and conductances of the coupling transconductors are absorbed in the four shunt RC networks.

It can be shown that the voltage of the secondary winding is given by

$$V_2 = \frac{G_{21}}{G_{m1}^2 G_{m2}}\frac{A(s)I_1 + B(s)I_2}{C(s)}, \tag{3.16}$$

where

$$A(s) = sC_{11} + G_{o11}, \tag{3.17}$$

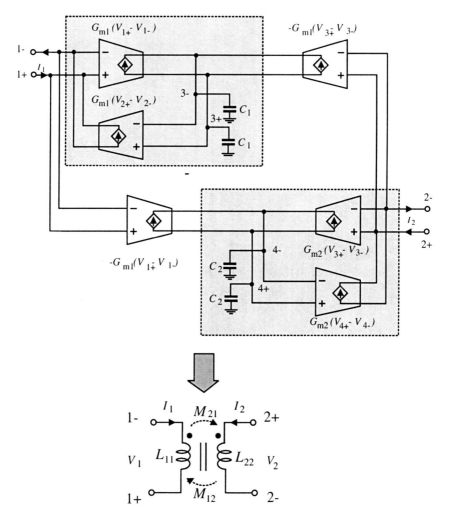

Figure 3.5. Lossless floating gyrator-C active transformers. The active inductors are coupled from their interface nodes 1 and 2 to the internal nodes 3 and 4.

$$B(s) = \frac{G_{m1}^2}{G_{21}G_{m2}}\left[\left(\frac{sC_{11} + G_{o11}}{G_{m1}}\right)\left(\frac{sC_{12} + G_{o12}}{G_{m1}}\right) + 1\right](sC_{22} + G_{o22}),$$

$$(3.18)$$

and

$$C(s) = \left[\left(\frac{sC_{11} + G_{o11}}{G_{m1}}\right)\left(\frac{sC_{12} + G_{o12}}{G_{m1}}\right) + 1\right]$$

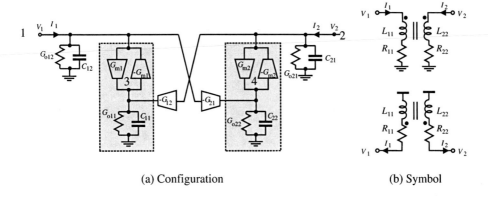

(a) Configuration   (b) Symbol

Figure 3.6. Lossy gyrator-C active transformers. C_{12} and G_{o12} represent the total capacitance and conductance encountered at node 1, respectively. They consist of the contribution of all the transconductors connected to the node. The same notation applies to the capacitance and conductance at other nodes of the transformer.

$$\left[\left(\frac{sC_{21} + G_{o21}}{G_{m2}}\right)\left(\frac{sC_{22} + G_{o22}}{G_{m2}}\right) + 1\right]$$
$$-\frac{G_{12}G_{21}}{G_{m1}G_{m2}}.$$
(3.19)

To simply (3.16), the following assumptions are made :

- $G_o \ll G_m$ holds for devices biased in the saturation. This assumption is generally valid.

- Active transformers are typically operated at frequencies below the cut-off frequencies of the transconductors synthesizing the transformers, which are given by

$$\omega_{t11,12} = \frac{G_{m1}}{C_{11,12}},$$
$$\omega_{t21,22} = \frac{G_{m2}}{C_{21,22}}.$$
(3.20)

Taking these into consideration, we have

$$\frac{sC_{11} + G_{o11}}{G_{m1}}, \frac{sC_{12} + G_{o12}}{G_{m1}}, \frac{sC_{21} + G_{o21}}{G_{m2}}, \frac{sC_{22} + G_{o22}}{G_{m2}} \approx 0.$$
(3.21)

Eqs.(3.18-3.19) are simplified to

$$B(s) \approx \frac{G_{m1}^2}{G_{21}G_{m2}}(sC_{22} + G_{o22}), \tag{3.22}$$

$$C(s) \approx \Delta. \tag{3.23}$$

The self impedance of the secondary winding of the transformer is obtained from

$$Z_{22} = \frac{V_2}{I_2}\bigg|_{I_1=0} = sL_{22} + R_{22}, \tag{3.24}$$

where

$$L_{22} = \frac{C_{22}}{G_{m2}^2\Delta}, \tag{3.25}$$

and

$$R_{22} = \frac{G_{o22}}{G_{m2}^2\Delta}. \tag{3.26}$$

Eq.(3.24) reveals that the secondary winding of the active transformer can be represented by an inductor of inductance L_{22} in series with a resistor of resistance R_{22}. R_{22} is termed the self-resistance of the secondary winding.

Following the similar argument, one can show that the transimpedance from the primary winding to the secondary winding is given by

$$Z_{21} = \frac{V_2}{I_1}\bigg|_{I_1=0} = sM_{21} + R_{21}, \tag{3.27}$$

where

$$M_{21} = \left(\frac{G_{21}}{G_{m2}}\right)\frac{C_{11}}{G_{m1}^2\Delta}, \tag{3.28}$$

is the mutual inductance from the primary winding to the secondary winding of the active transformer and

$$R_{21} = \left(\frac{G_{21}}{G_{m2}}\right)\frac{G_{o11}}{G_{m1}^2\Delta}. \tag{3.29}$$

R_{21} is the mutual resistance from the primary winding to the secondary winding of the active transformer. The self inductance and self resistance of the primary winding, the mutual inductance and resistance from the secondary winding to the primary winding of the active transformer can be obtained in a similar manner and the results are given below.

$$L_{11} = \frac{C_{11}}{G_{m1}^2 \Delta}, \tag{3.30}$$

$$R_{11} = \frac{G_{o11}}{G_{m1}^2 \Delta}, \tag{3.31}$$

$$M_{12} = \left(\frac{G_{12}}{G_{m1}}\right) \frac{C_{22}}{G_{m2}^2 \Delta}, \tag{3.32}$$

$$R_{12} = \left(\frac{G_{12}}{G_{m1}}\right) \frac{G_{o22}}{G_{m2}^2 \Delta}. \tag{3.33}$$

In what follows we examine these results in detail :

- The finite input and output resistances of the transconductors do not contribute to either self or mutual impedances. They, however, give rise to parasitic self and mutual resistances.

- Mutual inductances M_{12} and M_{21} are not the same. This differs from conventional transformers where only one mutual inductance exists. M_{12} and M_{21} are more preciously the trans-inductances from the secondary winding to the primary winding and that from the primary winding to the secondary winding, respectively. The same is true for R_{12} and R_{21}. Here we will still follow the conventional terminologies and term them mutual inductances and mutual resistances.

- The relations between the mutual inductances and self-inductances of active transformers are given by

$$M_{21} = \left(\frac{G_{21}}{G_{m2}}\right) L_{11}, \tag{3.34}$$

and

$$M_{12} = \left(\frac{G_{12}}{G_{m1}}\right)L_{22}. \qquad (3.35)$$

They are the same as the mutual inductances of lossless active transformers derived earlier.

- Mutual inductances M_{21} and M_{12} can be tuned by varying the transconductances of the coupling transconductors G_{12} and G_{21} without affecting the self inductances of the primary and secondary windings, provided that $\Delta \approx 1$ holds. This condition is the same as that of ideal active transformers.

- The relations between the mutual and self resistances of active transformers are given by

$$R_{21} = \left(\frac{G_{21}}{G_{m2}}\right)R_{11}, \qquad (3.36)$$

$$R_{12} = \left(\frac{G_{12}}{G_{m1}}\right)R_{22}. \qquad (3.37)$$

R_{12} and R_{21} can be lowered by reducing R_{11} and R_{22} without affecting the transconductances of the coupling transconductors. This enables the minimization of these unwanted parasitic resistances without affecting the self and mutual inductances of the transformers.

3.1.4   Active Transformers With Multiple Windings

The preceding gyrator-C active transformers have single primary winding and single secondary winding. We show in this section that the configurations of active transformers with single winding can be readily extended to active transformers with multiple windings. To simplify analysis, only lossless active transformers with multiple windings will be considered in this section. The configurations of lossy active transformers with multiple windings are the same as those of their lossless counterparts and their derivation is left as an exercise for readers.

A. Active Transformers with Multiple Primary Windings

The configuration of a lossless gyrator-C active transformer with multiple primary windings and single secondary winding is shown in Fig.3.7. The multiple primary windings are constructed using multiple active inductors $G_{m1} \sim C_1$, $G_{m2} \sim C_2$, ..., $G_{mN} \sim C_N$. They are coupled to the same secondary winding using a set of transconductors $G_{s,p1}$, $G_{s,p2}$, ..., $G_{s,pN}$. Write KCL at the node of the load capacitor C_s of the secondary winding of the transformer

$$sC_sV_{c,s} + G_{s,p1}V_{p1} + G_{s,p2}V_{p2} + G_{s,pN}V_{pN} - G_{ms}V_s = 0.$$

$$(3.38)$$

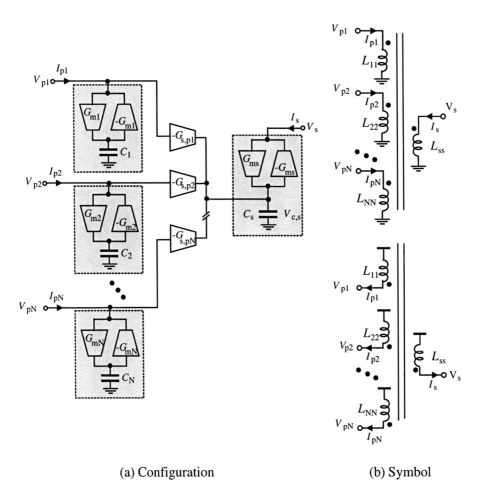

(a) Configuration   (b) Symbol

Figure 3.7. Unidirectional gyrator-C active transformers with multiple primary windings. The active inductors are coupled from the interface nodes of the primary windings to the internal node of the secondary windings.

Because

$$I_s = G_{ms} V_{c,s},$$

$$V_{p1} = \frac{sC_1}{G_{m1}^2} I_{p1},$$

$$V_{p2} = \frac{sC_2}{G_{m2}^2} I_{p2},$$

$$\ldots$$

$$V_{pN} = \frac{sC_N}{G_{mN}^2} I_{pN},$$

$$(3.39)$$

Eq.(3.38) becomes

$$
\begin{aligned}
V_s &= s\left(\frac{G_{s,p1}}{G_{ms}}\right)\frac{C_1}{G_{m1}^2} I_{p1} + s\left(\frac{G_{s,p2}}{G_{ms}}\right)\frac{C_2}{G_{m2}^2} I_{p2} + \ldots \\
&+ s\left(\frac{G_{s,pN}}{G_{ms}}\right)\frac{C_N}{G_{mN}^2} I_{pN} + s\left(\frac{C_s}{G_{ms}^2}\right) I_s \\
&= sM_{s1} I_{p1} + sM_{s2} I_{p2} + \ldots + sM_{sN} I_{pN} + sL_{ss} I_s.
\end{aligned}
$$

$$(3.40)$$

where

$$M_{s1} = \left(\frac{G_{s,p1}}{G_{ms}}\right)\frac{C_1}{G_{m1}^2},$$

$$M_{s2} = \left(\frac{G_{s,p2}}{G_{ms}}\right)\frac{C_2}{G_{m2}^2},$$

$$\ldots$$

$$M_{sN} = \left(\frac{G_{s,pN}}{G_{ms}}\right)\frac{C_N}{G_{mN}^2},$$

$$(3.41)$$

and

$$L_{ss} = \frac{C_s}{G_{ms}^2}.$$

$$(3.42)$$

It is evident from (3.40) that the voltage of the secondary winding of the active transformer is a linear function of the self inductance of the secondary winding and the mutual inductances from the primary windings to the secondary winding of the transformer.

B. Active Transformers with Multiple Secondary Windings

The configuration of gyrator-C active transformers with single primary winding and multiple secondary windings is shown in Fig.3.8. The multiple secondary windings are realized using multiple gyrator-C active inductors. The coupling from the primary winding to the secondary windings is established via a set of transconductors. The voltage of the secondary windings is given by

$$V_{s1} = s\left(\frac{G_{s1,p}}{G_{m,s1}}\right)\frac{C_p}{G_{mp}^2}I_p + \frac{sC_{s1}}{G_{ms1}^2}I_{s1} = sM_{s1,p}I_p + L_{ss1}I_{s1}.$$

$$V_{s2} = s\left(\frac{G_{s2,p}}{G_{m,s2}}\right)\frac{C_p}{G_{mp}^2}I_p + \frac{sC_{s2}}{G_{ms2}^2}I_{s2} = sM_{s2,p}I_p + L_{ss2}I_{s2},$$

$$\cdots$$

$$V_{sN} = s\left(\frac{G_{sN,p}}{G_{m,sN}}\right)\frac{C_p}{G_{mp}^2}I_p + \frac{sC_{sN}}{G_{msN}^2}I_{sN} = sM_{sN,p}I_p + L_{ssN}I_{sN},$$

where

$$M_{s1,p} = s\left(\frac{G_{s1,p}}{G_{m,s1}}\right)\frac{C_p}{G_{mp}^2},$$

$$L_{ss1} = \frac{sC_{s1}}{G_{ms1}^2},$$

$$M_{s2,p} = s\left(\frac{G_{s2,p}}{G_{m,s2}}\right)\frac{C_p}{G_{mp}^2},$$

(3.43)

$$L_{ss2} = \frac{sC_{s2}}{G_{ms2}^2},$$

$$\cdots$$

$$M_{sN,p} = s\left(\frac{G_{sN,p}}{G_{m,sN}}\right)\frac{C_p}{G_{mp}^2},$$

$$L_{ssN} = \frac{sC_{sN}}{G_{msN}^2}.$$

The voltage of each secondary winding is a linear function of both the self inductance of the secondary winding and the mutual inductance from the primary winding to the secondary winding.

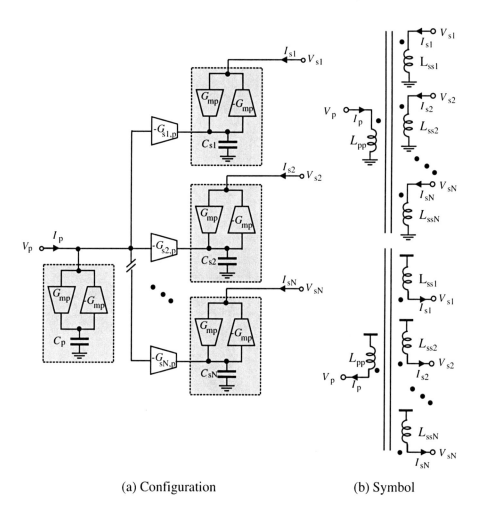

(a) Configuration   (b) Symbol

Figure 3.8. Unidirectional gyrator-C active transformers with multiple secondary windings. The active inductors are coupled from the interface node of the primary winding to the internal nodes of the secondary windings.

3.2   Characterization of Active Transformers

In this section, we investigate the most important figure-of-merits that quantify the performance of active transformers. These figure-of-merits include stability, frequency range, the tunability of self and mutual inductances, turn ratio, coupling factor, voltage and current transfer characteristics, impedance transformation, quality factor, noise, linearity, supply voltage sensitivity, parameter sensitivity, signal sensitivity, and power consumption.

3.2.1 Stability

The stability of active transformers can be investigated using (3.16). To simplify analysis, we assume

$$
\begin{aligned}
&1) \quad G_{o11} = G_{o12} = G_{o21} = G_{o22} = G_o, \\
&2) \quad G_{m1} = G_{m2} = G_m, \\
&3) \quad G_{12} = G_{21} = G_c, \\
&4) \quad C_{11} = C_{12} = C_{21} = C_{22} = C.
\end{aligned}
\tag{3.44}
$$

Because active transformers are typically operated at frequencies below the cutoff frequency of the transconductors constituting the active transformers, $\left(\dfrac{\omega}{\omega_t}\right)^2 \approx 0$ holds, where $\omega_t = \dfrac{G_m}{C}$ is the cut-off frequency of the transconductors. Further because $G_o < G_m$ for devices biased in the saturation, $\left(\dfrac{G_o}{G_m}\right)^2 \approx 0$ follows. Eq.(3.19) is simplified to

$$
C \approx s^2 \left(\frac{4 G_o^2}{\omega_t^2 G_m^2}\right) + s\left(\frac{4 G_o}{\omega_t G_m}\right) + \Delta.
\tag{3.45}
$$

The poles of V_2 are given by

$$
\begin{aligned}
s_1 &= \frac{\omega_t g_m}{2 G_o}\left(-1 + \frac{G_c}{G_m}\right), \\
s_2 &= \frac{\omega_t g_m}{2 G_o}\left(-1 - \frac{G_c}{G_m}\right).
\end{aligned}
\tag{3.46}
$$

It is seen from (3.46) that in order to ensure that the system is stable, $G_c < G_m$ is required. This requirement is satisfied when $\Delta > 0$ holds. Note that $G_c < G_m$ is the condition to ensure that the values of both the self and mutual inductances of active transformers are positive.

3.2.2 Frequency Range

It was shown earlier that a lossless gyrator-C active transformer exhibits a mutual inductive characteristic between its two windings across the entire frequency spectrum. A lossy gyrator-C active transformer, however, only exhibits a mutual inductive characteristic between its two windings over a specific frequency range. The same is true for the self-inductances of the transformer as well. The frequency range of the self-inductance of the primary winding and that of the secondary winding are the same as gyrator-C active inductors and were investigated in Chapter 2. The frequency range over which a mutual inductive characteristic exists can be estimated by examining the mutual impedance of the transformer. It is seen from (3.16) that the mutual impedance from the primary winding to the secondary winding of the transformer is given by

$$
\begin{aligned}
Z_{21} &= \frac{A(s)}{C(s)} \\
&= \frac{sC + G_o}{s^2 \left(\dfrac{4G_o^2}{\omega_t^2 G_m^2} \right) + s \left(\dfrac{4G_o}{\omega_t G_m} \right) + \Delta}.
\end{aligned}
\tag{3.47}
$$

To simplify the analysis, the assumptions given in (3.44) are utilized. The frequency of the zero of Z_{21} is determined from $A(s) = 0$

$$
\omega_z = \frac{G_o}{C}.
\tag{3.48}
$$

The pole resonant frequency of Z_{21} is obtained from $C(s) = 0$. From (3.45), we have

$$
s^2 + s \left(\frac{G_o}{\omega_t G_m} \right) + \frac{\omega_t^2 G_m^2}{4G_o^2} \Delta = 0.
\tag{3.49}
$$

The pole resonant frequency of Z_{21} is given by

$$
\omega_p \approx \left(\frac{G_m}{2G_o} \right) \omega_t,
\tag{3.50}
$$

where we have further assumed

$$
\Delta = 1 - \left(\frac{G_c}{G_m} \right)^2 \approx 1
\tag{3.51}
$$

in simplification of (3.50). A mutual inductive characteristic from the primary winding to the secondary winding of the active transformer exists when $\omega_z < \omega < \omega_p$, as shown in Fig.3.9 graphically.

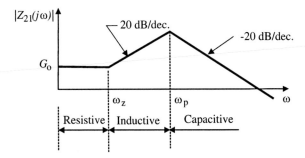

Figure 3.9. Frequency range of active transformers.

3.2.3 Tunability of Self and Mutual Inductances

Applications such as frequency selective networks (filters) usually require that the inductances of active transformers, both self and mutual inductances, be tunable with large inductance tuning ranges. Similar to the inductance tuning of active inductors, the self inductances of a gyrator-C active transformer can be tuned by either changing the load capacitance or the transconductances of the transconductors of respective winding. It should be noted that a variation of the self-inductances of an active transformer will affect corresponding mutual inductances. The mutual inductances of active transformers can be tuned by varying the transconductances of the coupling transconductors of the transformer without affecting the self inductances of the transformer. Further, conductance tuning can be used for the coarse tuning of the self inductances of the transformer due to its large tuning range while capacitance tuning can be used for the fine tuning of the self inductances of the transformer due to its small tuning range.

3.2.4 Turn ratios

The turn ratios from the primary winding to the secondary winding of an ideal transformer are defined as

$$n_{21} = \sqrt{\frac{L_{22}}{L_{11}}},$$

$$n_{12} = \sqrt{\frac{L_{11}}{L_{22}}}. \tag{3.52}$$

Substituting (3.30) and (3.25) into (3.52), we obtain the turn ratios of active transformers

$$n_{21} = \left(\frac{G_{m1}}{G_{m2}}\right)\sqrt{\frac{C_2}{C_1}},$$

$$n_{12} = \left(\frac{G_{m2}}{G_{m1}}\right)\sqrt{\frac{C_1}{C_2}},$$
(3.53)

It is seen from (3.53) that the transconductances of the coupling transconductors G_{12} and G_{21} have no effect on the turn ratios of the transformer. The finite input and output resistances of the transconductors constituting the active transformer also do not affect the turn ratios. The turn ratios are completely determined by the property of the two winding active inductors. Also, the turn ratios are more sensitive to the ratio of the transconductances of the transconductors than the ratio of the load capacitances of the transconductors.

3.2.5   Coupling Factors

The coupling factors between the primary winding and the secondary winding of a transformer are defined as

$$k_{21} = \frac{M_{21}}{\sqrt{L_{11}L_{22}}},$$

$$k_{12} = \frac{M_{12}}{\sqrt{L_{11}L_{22}}}.$$
(3.54)

Substituting the expressions of the self and mutual inductances of active transformers into (3.54), we obtain the coupling factors of active transformers

$$k_{21} = \left(\frac{G_{21}}{G_{m1}}\right)\sqrt{\frac{C_1}{C_2}},$$

$$k_{12} = \left(\frac{G_{12}}{G_{m2}}\right)\sqrt{\frac{C_2}{C_1}}.$$
(3.55)

It is seen from (3.55) that the coupling factors are directly proportional to the transconductances of the coupling transconductors. They can be tuned by varying G_{12} and G_{21} without affecting the self inductances of the transformer. The mutual inductances, however, will change accordingly.

3.2.6   Voltage Transfer Characteristics

The voltage transfer characteristics from the primary winding to the secondary winding of an active transformer can be derived by applying an ideal voltage source v_1 to the primary winding of the transformer while leaving the secondary winding of the transformer open-circuited, as shown in Fig.3.10. Because $I_2 = 0$, we have

$$V_2 = sM_{21}I_1, \tag{3.56}$$

and

$$V_1 = sL_{11}I_1. \tag{3.57}$$

The ratio of (3.56) to (3.57) yields

$$\begin{aligned} V_2 &= \left(\frac{M_{21}}{L_{11}}\right)V_1 \\ &= \left(\frac{G_{21}}{G_{m2}}\right)V_1. \end{aligned} \tag{3.58}$$

The preceding results reveal that the voltage transfer characteristics from the primary winding to the secondary winding of the active transformer can be changed by varying the degree of the coupling from the primary winding active inductor to the secondary winding active inductor of the transformer. This differs from the voltage transfer characteristics of ideal transformers, which are given by

$$V_2 = n_{21}V_1. \tag{3.59}$$

The voltage transfer characteristics of ideal transformers are solely determined by the turn ratio of the transformers. The reason for this difference is that in the case of ideal transformers, there is a 100% magnetic coupling between the two windings of the transformers. The coupling of active transformers, however, is determined by the transconductances of the coupling transconductors that link the two active inductors of the transformers.

The voltage transfer characteristics of an active transformer can also be depicted using the coupling factors of the active transformer. Eq.(3.58) can be written as

$$\begin{aligned} V_2 &= \left(\frac{G_{21}}{G_{m2}}\right)V_1 \\ &= k_{21}\left(\frac{G_{m1}}{G_{m2}}\sqrt{\frac{C_2}{C_1}}\right)V_1 \\ &= k_{21}(n_{21}V_1). \end{aligned} \tag{3.60}$$

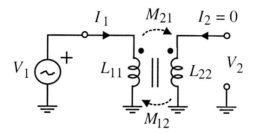

Figure 3.10. Determination of the voltage transfer characteristics of active transformers.

Eq.(3.60) reveals both the similarity and distinction between the voltage transfer characteristics of active transformers and those of ideal transformers. Unlike ideal transformers with a perfect coupling ($k_{21} = 1$) whose voltage transfer characteristics are given by (3.59) [103], the voltage transfer characteristics of active transformers are scaled by both the turn ratio and the coupling coefficient of the transformers.

3.2.7   Current Transfer Characteristics

To derive the current transfer characteristics from the primary winding to the secondary winding of active transformers, consider the lossless active transformer shown in Fig.3.11 where a test voltage source V_1 is applied to the primary winding of the transformer while the secondary winding of the transformer is short-circuited. Because $V_2 = 0$, we have

$$
\begin{aligned}
I_2 &= -\frac{M_{21}}{L_{22}}I_1 \\
&= -\frac{G_{21}}{G_{m2}}\frac{L_{11}}{L_{22}}I_1.
\end{aligned}
\tag{3.61}
$$

Making use of the expression of the turn ratio, Eq.(3.61) becomes

$$
I_2 = -\frac{G_{21}}{G_{m2}}\frac{1}{n_{21}^2}I_1.
\tag{3.62}
$$

Further substituting the expression of the coupling factor into (3.62) yields

$$
\begin{aligned}
I_2 &= -k_{21}\left(\frac{g_{m1}}{g_{m2}}\sqrt{\frac{C_2}{C_1}}\frac{1}{n_{21}^2}\right)I_1 \\
&= -k_{21}\left(\frac{I_1}{n_{21}}\right).
\end{aligned}
\tag{3.63}
$$

Eq.(3.63) shows that unlike ideal transformers with a perfect coupling ($k_{21} = 1$) whose current transfer characteristics are given by [103]

$$I_2 = -\frac{I_1}{n_{21}}, \tag{3.64}$$

the current transfer characteristics of active transformers are scaled by both the turn ratios and the coupling coefficients of the transformers.

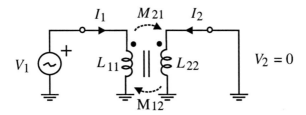

Figure 3.11. Derivation of the current transfer characteristics of active transformers.

3.2.8 Impedance Transformation

The terminal behavior of the two-port network shown in Fig.3.12 is typically depicted by its impedance parameters

$$V_1 = Z_{11}I_1 + Z_{12}I_2,$$
$$V_2 = Z_{21}I_1 + Z_{22}I_2, \tag{3.65}$$

where Z_{11} and Z_{22} are the input impedances at ports 1 and 2, respectively, and Z_{12} and Z_{21} are the transimpedance from port 2 to port 1 and that from port 1 to port 2 of the two-port network, respectively.

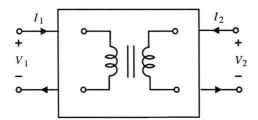

Figure 3.12. Two-port network representation of active transformers.

Active transformers are two-port networks with port 1 the primary winding and port 2 the secondary winding. The self-impedance of the primary winding and that of the secondary winding, denoted by Z_{11} and Z_{22}, respectively, the mutual inductance from the secondary winding to the primary winding and that from the primary winding to the secondary winding, denoted by Z_{12} and Z_{21} respectively, of an active transformer are obtained from

$$Z_{11} = \left.\frac{V_1}{I_1}\right|_{I_2=0},$$

$$Z_{22} = \left.\frac{V_2}{I_2}\right|_{I_1=0},$$

$$Z_{12} = \left.\frac{V_1}{I_2}\right|_{I_1=0}, \qquad\qquad (3.66)$$

$$Z_{21} = \left.\frac{V_2}{I_1}\right|_{I_2=0}.$$

For lossless active transformers, because

$$Z_{22} = sL_{22} = \frac{sC_{22}}{G_{m2}^2\Delta},$$

$$Z_{11} = sL_{11} = \frac{sC_{11}}{G_{m1}^2\Delta},$$

$$Z_{12} = sM_{12} = s\left(\frac{G_{12}}{G_{m1}}\right)\frac{C_{22}}{G_{m2}^2\Delta}, \qquad\qquad (3.67)$$

$$Z_{21} = sM_{12} = s\left(\frac{G_{21}}{G_{m2}}\right)\frac{C_{11}}{G_{m1}^2\Delta},$$

we have

$$Z_{22} = n_{21}^2 Z_{11}, \qquad\qquad (3.68)$$

and

$$Z_{12} = \left(\frac{k_{12}}{k_{21}}\right)Z_{21}. \qquad\qquad (3.69)$$

Eq.(3.68) bears a strong resemblance to the impedance transformation characteristics of passive lossless transformers whereas (3.69) is unique to active transformers.

3.2.9 Noise

The effect of the noise sources of an active transformer at the secondary winding terminal of the transformer can be represented by a noise-voltage generator $\overline{v_n^2}$ and a noise-current generator $\overline{i_n^2}$ at the primary winding terminal of the active transformer, as shown in Fig.3.13. To simplify analysis, we assume that each active inductor is made of two identical transconductors and a load capacitor. The noise of each transconductor is represented by its input-referred noise-voltage generator. Also, all noise sources are assumed to be uncorrelated.

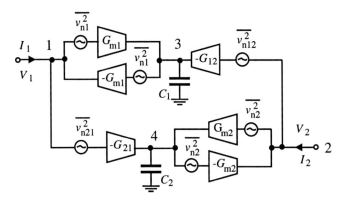

(a) Active transformers with noise sources

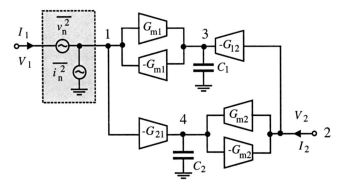

(b) Input-referred noise-vltage and
noise-current generators

Figure 3.13. Noise analysis of gyrator-C active transformers.

To obtain $\overline{v_n^2}$, port 1 of the active transformer shown in Fig.3.13(a) is short-circuited and the power of the noise at the output of the secondary winding of the transformer is derived using elementary noise analysis

$$\overline{v_{n,2}^2} = \left(\frac{G_{21}}{G_{m2}}\right)^2 \overline{v_{n21}^2} + \left[1 + \left(\frac{\omega}{\omega_{t2}}\right)^2\right]\overline{v_{n2}^2}, \tag{3.70}$$

where $\omega_{t2} = \dfrac{G_{m2}}{C_2}$. The output noise power of Fig.3.13(b) with only $\overline{v_n^2}$ applied to the input port is then obtained.

$$\overline{v_{n,2}^2} = \left(\frac{G_{21}}{G_{m2}}\right)^2 \overline{v_n^2}. \tag{3.71}$$

Equating (3.70) and (3.71) yields

$$\overline{v_n^2} = \overline{v_{n21}^2} + \left(\frac{G_{m2}}{G_{21}}\right)^2\left[1 + \left(\frac{\omega}{\omega_{t2}}\right)^2\right]\overline{v_{n2}^2}. \tag{3.72}$$

$$\tag{3.73}$$

It is seen from the preceding analysis that when port 1 of Fig.3.13(a) is short-circuited, v_{n1} will not contribute to the total noise at node 2 as the transconductor G_{12} is directional. Because the input impedance of G_{12} is infinite and node 1 is grounded, v_{n12} has no impact on the noise at node 2.

To derive $\overline{i_n^2}$, the input port of the active transformer in Fig.3.13(a) is open-circuited and the noise power at node 2 is derived

$$\begin{aligned}\overline{v_{n,2}^2} &= \frac{1}{\Delta}\left\{\left(\frac{G_{21}}{G_{m2}}\right)^2\left[1 + \left(\frac{\omega}{\omega_{t1}}\right)^2\right]\overline{v_{n1}^2} + \left[1 + \left(\frac{\omega}{\omega_{t2}}\right)^2\right]\overline{v_{n2}^2} \right. \\ &\quad \left. + \left(\frac{G_{12}G_{21}}{G_{m1}G_{m2}}\right)^2\overline{v_{n12}^2} + \left(\frac{G_{21}}{G_{m2}}\right)^2\overline{v_{n21}^2}\right\}. \end{aligned} \tag{3.74}$$

The output noise power of Fig.3.13(b) with i_n injected at the input port is given by

$$\overline{v_{n,2}^2} = \frac{G_{21}^2}{G_{m1}^2 g_{m2}^2}\left(\frac{\omega}{\omega_{t1}}\right)^2\frac{\overline{i_n^2}}{\Delta}. \tag{3.75}$$

Equating (3.74) and (3.75) yields

$$\overline{i_n^2} = \left(\frac{\omega_{t2}}{\omega}\right)^2 \left[\left(\frac{G_{m1}^4}{G_{m2}^2}\right)\overline{v_{n1}^2} + \left(\frac{G_{m1}^4}{G_{21}^2}\right)\overline{v_{n2}^2} + \left(\frac{G_{m1}^4}{G_{m2}^2}\right)\overline{v_{n21}^2}\right.$$

$$+ \left.\left(\frac{G_{m1}}{G_{m2}}\right)^2 G_{12}^2 \overline{v_{n12}^2}\right], \tag{3.76}$$

$$\tag{3.77}$$

where $\omega \ll \omega_{t2}$ was utilized to simply analysis.

3.2.10 Quality Factors

There are four inductances associated with a lossless active transformer, namely self-inductances L_{11} and L_{22}, and mutual inductances M_{12} and M_{21}. For a lossy active transformer, there are four parasitic resistances, namely self-resistances R_{11} and R_{22}, and mutual resistances R_{12} and R_{21} that quantify the ohmic losses of the transformer. Four corresponding quality factor quantities, namely the self quality factor of the primary winding, denoted by Q_{11}, the self quality factor of the secondary winding, denoted by Q_{22}, the mutual quality factor from the primary winding to the secondary, denoted by Q_{21}, and the mutual quality factor from the secondary winding to the primary winding, denoted by Q_{12}, quantify the ratio of the energy stored in the transformer to the energy dissipated by the transformer, exist.

It was shown in Chapter 2 that the quality factor of an active inductor at frequencies below the cutoff frequency of the transconductors synthesizing the active inductor can be approximated by

$$Q \approx \frac{\omega L}{R_s} \qquad (R_s \text{ dominates}). \tag{3.78}$$

Eq.(3.78) can be used to quantify the four preceding quality factors of the active transformer.

$$Q_{11} = \frac{\omega L_{11}}{R_{11}} = \frac{\omega C_{11}}{G_{o11}}, \tag{3.79}$$

$$Q_{22} = \frac{\omega L_{22}}{R_{22}} = \frac{\omega C_{22}}{G_{o22}}, \tag{3.80}$$

$$Q_{21} = \frac{\omega M_{21}}{R_{21}} = \frac{\omega C_{11}}{G_{o11}}, \tag{3.81}$$

$$Q_{12} = \frac{\omega L_{12}}{R_{12}} = \frac{\omega C_{22}}{G_{o22}}. \tag{3.82}$$

We comment on the preceding results :

- Self quality factors Q_{11} and Q_{22} quantify the ratios of the energy stored in the primary and secondary windings to the energy loss in the primary and secondary windings of active transformers, respectively. Note that the energy storage is due to C_{11} and C_{22} while the energy loss is due to G_{o11} and G_{o22}.

- Mutual quality factors Q_{21} and Q_{12} quantify the ratios of the energy transferred between the two windings to the energy loss during the transfer due to the lossy coupling of the two windings of active transformers. Note $Q_{11} = Q_{21}$ and $Q_{22} = Q_{12}$.

- G_{o21} and G_{o12} can be considered as a part of the external circuitry connected to the active transformer and therefore do not contribute to the quality factors of the transformers. To improve the quality factors of active transformers, only G_{o11} and G_{o22} need to be minimized.

- The quality factors of active transformers are bias-dependent and swing-dependent. This is because G_{o11} and G_{o22}, which are the results of both the input and output conductances of the transconductors of active transformers, are bias-dependent and swing-dependent. Two cases exist:

  - Case 1 - The input conductances of the transconductors of active transformers are infinite. In this case, G_{o11} and G_{o22} are only due to the output conductances of the transconductors of active transformers. When the transistors of active transformers are biased in the saturation region, G_{o11} and G_{o22} are small. However, when the devices are biased in the triode region, G_{o11} and G_{o22} become large. Further when the devices are operated in the transition region between the triode and saturation regions, G_{o11} and G_{o22} are no longer constant and vary with the swing of the voltages of the transformers.

  - Case 2 - G_{o11} or G_{o22} are dominated by the input conductances of the transconductors of active transformers. An typical example is the case where the transconductor is common-gate configured. In this case, the input resistance of the common-gate configured transconductor is given by $\frac{1}{g_m}$, which is strongly dependent of the dc biasing condition.

The quality factors of spiral transformers, on the other hand, are constant at a given frequency and do not vary with signal swing. This bias/swing-dependent characteristic of active transformers imposes design challenges when active transformers are used in applications where either the biasing conditions or the signal swing of the transformers experience large variations.

3.2.11 Linearity

The preceding development of gyrator-C active transformers assumes that the transconductors of the transformers are linear. This assumption is only valid if the swing of the input voltage of the transconductors of the transformers is small. When the voltage swing is large, the transconductors will exhibit a nonlinear characteristic and the synthesized active transformers are no longer linear.

The linearity constraint of the transconductors sets the maximum swing of the voltage of active transformers for a given level of distortion. If we assume that the transconductances of the transistors of active transformers are constant when the transistors are biased in the saturation, then the maximum swing of the voltage of the active transformers can be estimated from the pinch-off condition of the transistors. When the transistors enter the triode region, the transconductances of the transistors will decrease from g_n (saturation) to g_{ds} (triode) in a nonlinear fashion. The self-inductance of the primary winding of the active transformer will increase from $L_{11} = \dfrac{C_{11}}{G_{m1}^2}$ (saturation) to $L_{11} = \dfrac{C_{11}}{G_{ds1}^2}$ (triode) accordingly. Other inductances of the transformer will also change in a similar way. Further, the parasitic resistances of the active transformer $R_{11}, R_{12}, R_{21},$ and R_{22} will also change due to the variation of $G_{m1}, G_{m2}, G_{o11},$ and G_{o22}.

3.2.12 Supply Voltage Sensitivity

The supply voltage sensitivity of the inductances of active transformers quantifies the effect of the variation of the supply voltage on the inductances of the active transformers. The fluctuation of the supply voltage V_{DD} affects the inductances of an active transformer mainly by altering the dc operating points of the transconductors subsequently the transconductances of the transconductors of the active transformer.

Assume that the load capacitance C of an active inductor does not vary with V_{DD}. Further assume that $G_{m1}G_{m2} \gg G_{12}G_{21}$, i.e. $\Delta \approx 1$. The supply voltage sensitivity of the self inductance L_{11} of the active transformer is obtained from differentiating L_{11} of (3.30) with respect to V_{DD}

$$\frac{\partial L_{11}}{\partial V_{DD}} = -\frac{2C_1}{G_{m1}^3}\frac{\partial G_{m1}}{\partial V_{DD}}. \tag{3.83}$$

The normalized sensitivity is obtained from

$$S_{V_{DD}}^{L_{11}} = \frac{V_{DD}}{L_{11}} \frac{\partial L_{11}}{\partial V_{DD}} = -2S_{V_{DD}}^{G_{m1}}. \tag{3.84}$$

Similarly, one can show that

$$S_{V_{DD}}^{L_{22}} = -2S_{V_{DD}}^{G_{m2}},$$

$$S_{V_{DD}}^{M_{12}} = S_{V_{DD}}^{G_{12}} - 2S_{V_{DD}}^{G_{m2}} - S_{V_{DD}}^{G_{m1}}, \tag{3.85}$$

$$S_{V_{DD}}^{M_{21}} = S_{V_{DD}}^{G_{21}} - 2S_{V_{DD}}^{G_{m1}} - S_{V_{DD}}^{G_{m2}}.$$

where $S_{V_{DD}}^{G_{m1}}$, $S_{V_{DD}}^{G_{m2}}$, $S_{V_{DD}}^{G_{12}}$, and $S_{V_{DD}}^{G_{21}}$ are the normalized supply voltage sensitivities of G_{m1}, G_{m2}, G_{12}, and G_{21}, respectively. The preceding analysis reveals that to minimize the effect of V_{DD} fluctuation on the performance of active transformers, the transconductances of the transconductors of the active transformers should be constant.

3.2.13   Parameter Sensitivity

Similar to the parameter sensitivity of active inductors, the performance of active transformers is sensitive to parameter spread. The sensitivity of the inductance of an active transformer to parameter x_j of the transformer defined as

$$D_{x_j}^{L} = \frac{\partial L}{\partial x_j} \tag{3.86}$$

quantifies the effect of the variation of parameter x_j on the inductance of the active transformer. By assuming that the parameters of the active transformer are Gaussian distributed and uncorrelated, the overall effect of the variation of the parameters of the active transformer on the inductance of the transformer is obtained from

$$\sigma_L^2 = \sum_{j=1}^{N} \left(\frac{\partial L}{\partial x_j} \right)^2 \sigma_{x_j}^2, \tag{3.87}$$

where σ_L and σ_{x_j} denote the standard deviations of L and x_j, respectively, and N is the number of the parameters of the active transformer. To compute the parameter sensitivity of L_{11} of an active transformer, because

$$\frac{\partial L_{11}}{\partial C_1} = \frac{1}{G_{m1}^2 \Delta},$$

$$\frac{\partial L_{11}}{\partial G_{m1}} = -\frac{2C_1}{G_{m1}^3 \Delta}, \qquad (3.88)$$

we have

$$\frac{\sigma_{L_{11}}^2}{L_{11}^2} = \frac{\sigma_{C_1}^2}{C_1^2} + 4\frac{\sigma_{G_{m1}}^2}{G_{m1}^2}. \qquad (3.89)$$

In a similar manner, one can show that

$$\frac{\sigma_{L_{22}}^2}{L_{22}^2} = \frac{\sigma_{C_2}^2}{C_2^2} + 4\frac{\sigma_{G_{m2}}^2}{G_{m2}^2}.$$

$$\frac{\sigma_{M_{12}}^2}{M_{12}^2} = \frac{\sigma_{G_{12}}^2}{G_{12}^2} + \frac{\sigma_{G_{m1}}^2}{G_{m1}^2} + \frac{\sigma_{C_2}^2}{C_2^2} + 4\frac{\sigma_{G_{m2}}^2}{G_{m2}^2}. \qquad (3.90)$$

$$\frac{\sigma_{M_{21}}^2}{M_{21}^2} = \frac{\sigma_{G_{21}}^2}{G_{21}^2} + \frac{\sigma_{G_{m2}}^2}{G_{m2}^2} + \frac{\sigma_{C_1}^2}{C_1^2} + 4\frac{\sigma_{G_{m1}}^2}{G_{m1}^2}.$$

Similar to active inductors, corner analysis and Monte Carlo analysis are the most widely analysis methods to quantify the effect of parameter spread on the performance of active transformers. The former determines the inductances of active transformers at process corners while the later quantifies the degree of the spread of the inductances of active transformers. Both analyses are required prior to any fabrication to ensure that the designed active transformers meet design specifications in all process conditions.

3.2.14 Power Consumption

The power consumption of an active transformer consists of three components : the power consumption of the primary winding active inductor, the power consumption of the secondary winding active transformer, and the power consumption of the coupling transconductors. The total power consumption of an active transformer can be estimated from

$$P_{total} = (J_p + J_s + J_c)V_{DD}, \qquad (3.91)$$

where J_p, J_s, and J_c are the total currents drawn by the primary winding, the secondary winding, and the coupling network of the active transformer, respectively.

It was pointed out in Chapter 2 that the power consumption of gyrator-C active inductors is usually not of a critical concern because the inductance of these inductors is inversely proportional to the transconductances of the transconductors. To have a large inductance, G_{m1} and G_{m2} are usually made small by lowering the dc biasing current of the transconductors. The static power consumption of the primary and secondary windings can be analyzed in a similar way as that of active inductors. The power consumption of the coupling transconductors is typically low simply because these coupling transconductors are very simple and their transconductances must be smaller as compared with the transconductances of the primary and secondary windings in order to satisfy $\Delta < 1$. The power consumption of an active transformer is typically dominated by its compensating negative resistor network.

3.3 Implementation of Active Transformers

An active transformer is constructed by coupled two active inductors using transconductor networks. For each CMOS active inductor studied in Chapter 2, a corresponding set of active transformers can be constructed in this way. Similar to the implementation of CMOS active inductors, the transconductors of active transformers should be configured as simple as possible to maximize the self-resonant frequency and to minimize the silicon area and power consumption of the transformers.

3.3.1 Basic Active Transformers

The active transformers shown in Figs.3.14, 3.15, 3.16, and 3.17 are evolved from the basic active inductors investigated in Chapter 2 [104]. The coupling transconductors between the primary and secondary windings of the transformers are only a pair of MOS transistors. The coupling is from the interface nodes of the active inductor of one winding of the transformer to the internal nodes of the active inductor of the other winding of the transformer. There is no direct current path between the two windings of the transformer such that each winding can be biased and tuned individually. In Figs.3.14 and 3.15, all transistors of the active transformers are isolated from the ground rail by two biasing current sources $J_{2,4}$ whereas in Figs.3.16 and 3.17, all transistors of the active transformers are isolated from the supply voltage rail by two biasing current sources $J_{1,3}$. These distinct characteristics are reflected by the symbols of the active transformers, also shown in the figures.

The direction of the coupling of the transformers identified by the location of the two dots in the symbols in Figs.3.14 and 3.15 is determined as the followings :

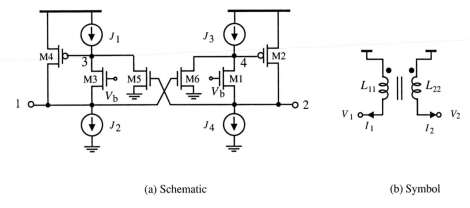

(a) Schematic   (b) Symbol

Figure 3.14. Simplified schematic of single-ended basic gyrator-C active transformers. Active inductors are coupled from the interface nodes to the internal nodes.

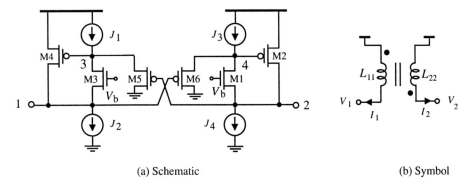

(a) Schematic   (b) Symbol

Figure 3.15. Simplified schematic of single-ended basic gyrator-C active transformers. Active inductors are coupled from the interface nodes to the internal nodes.

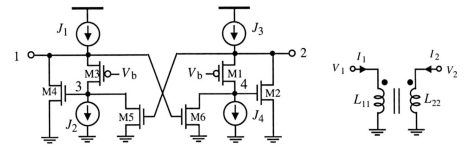

Figure 3.16. Simplified schematic of single-ended basic gyrator-C active transformers. Active inductors are coupled from the interface nodes to the internal nodes.

- nMOS-coupled - An increase of v_1 will result in an increase of i_{D6}. As a result, a smaller current will be injected into node 4. The coupling transconductor thus has a negative transconductance.

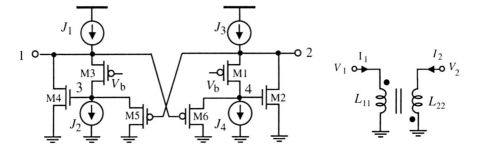

Figure 3.17. Simplified schematic of single-ended basic gyrator-C active transformers. Active inductors are coupled from the interface nodes to the internal nodes.

- pMOS-coupled - An increase of v_1 will result in a decrease of i_{D6}. This is echoed with an increase in the current injected into node 4. The transconductance of the coupling transconductor is positive.

The direction of the coupling of the transformers in Figs.3.16 and 3.17 can also be determined in a similar way and is shown in the figure.

The parameters of the transconductors of the active transformers, in particular the transconductances and load capacitance of the active inductors of the transformer, are tabulated in Table 3.1 where g_{ds} and the parasitics of MOSFETs are neglected, the capacitances and g_o of the coupling networks are neglected, and $g_{m1} = g_{m2} = g_{m1,2}, g_{m3} = g_{m4} = g_{m3,4}$ are assumed. Table 3.2 tabulates the inductances and parasitic resistances of the active transformers. The voltage swing at the input node of the basic active transformers is estimated from the pinch-off condition of MOSFETs and the results are tabulated in Table 3.3.

Table 3.1. Parameters of single-ended basic active transformers.

Blocks	Load cap.	Conductances	Transconductances
Primary winding active inductor	$C_{12} \approx C_{gs3}$ $C_{11} = C_{gs4}$	$G_{o12} \approx g_{m3}$ $G_{o11} = g_{o3}$	$G_{m1} = g_{m3,4}$ $G_{m2} = g_{m1,2}$
Secondary winding active inductor	$C_{21} \approx C_{gs1}$ $C_{22} = C_{gs2}$	$G_{o21} \approx g_{m1}$ $G_{o22} = g_{o1}$	$G_{m1} = g_{m3,4}$ $G_{m2} = g_{m1,2}$
Coupling transconductor			$G_{21} = g_{m6}$ $G_{12} = g_{m5}$

It is seen that the active transformers of Fig.3.14 and Fig.3.15 have very small voltage swing ranges. Biasing of these transformers becomes very difficult.

Table 3.2. Inductances and parasitic resistances of single-ended basic active transformers ($\Delta = 1 - \dfrac{g_{m5}g_{m6}}{g_{m1,2}g_{m3,4}}$).

Windings	Inductances	Parasitic resistances
Primary	$L_{11} = \dfrac{C_{gs4}}{g_{m3,4}^2 \Delta}$	$R_{11} = \dfrac{g_{o3}}{g_{m3,4}^2 \Delta}$
Secondary/Primary	$M_{12} = \left(\dfrac{g_{m5}}{g_{m3,4}}\right)\dfrac{C_{gs2}}{g_{m1,2}^2 \Delta}$	$R_{12} = \left(\dfrac{g_{m5}}{g_{m3,4}}\right)\dfrac{g_{o1}}{g_{m1,2}^2 \Delta}$
Primary/Secondary	$M_{21} = \left(\dfrac{g_{m6}}{g_{m1,2}}\right)\dfrac{C_{gs4}}{g_{m3,4}^2 \Delta}$	$R_{21} = \left(\dfrac{g_{m6}}{g_{m1,2}}\right)\dfrac{g_{o3}}{g_{m3,4}^2 \Delta}$
Secondary	$L_{22} = \dfrac{C_{gs2}}{g_{m1,2}^2 \Delta}$	$R_{22} = \dfrac{g_{o1}}{g_{m1,2}^2 \Delta}$

Table 3.3. The maximum swing of the input voltage of single-ended basic transformers.

Transformers	Min. voltage	Max. voltage
Fig.3.14	$V_{min} = V_T$	$V_{max} = V_{DD} - V_T - V_{sat}$
Fig.3.15	$V_{min} = V_{sat}$	$V_{max} = V_{DD} - 2V_T$
Fig.3.16	$V_{min} = V_T$	$V_{max} = V_{DD} - V_{sat}$
Fig.3.17	$V_{min} = V_T + 2V_{sat}$	$V_{max} = V_{DD} - V_T$

The active transformers of Fig.3.16 and Fig.3.17, in comparison, have large voltage swing ranges and their proper dc biasing can be attained without a difficulty. The preceding observations also reveal that although theoretically an active transformer can be constructed by coupling two active inductors via a coupling network, the voltage swing and proper biasing requirements often disqualify some of these configurations.

3.3.2 Tang Active Transformers

The active transformers shown in Figs.3.18 and 3.19 were proposed by Tang *et al.* [105]. They are constructed from Wu current-reuse active inductors investigated in Chapter 2. Both the primary and secondary windings of the transformers are Wu current reuse active inductors. The coupling transconductors are two MOS transistors. The coupling between the primary and secondary windings is from the interface nodes to the internal nodes of the active inductors.

The direction of the coupling is determined as the followings : An increase of v_1 of the active transformer of Fig.3.18 will result in an increase in i_{D6}. This is echoed with an increasing current flowing out of node 4. The coupling transconductor thus has a negative transconductance. In a similar manner, one can show that the coupling transconductor of the active transformer of Fig.3.18 also has a negative transconductance. The maximum swing of the input voltage of the transformers is given in Table 3.4.

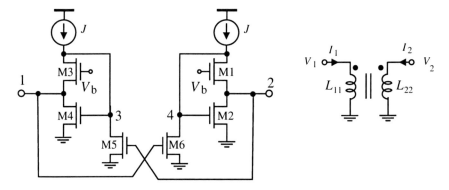

Figure 3.18. Simplified schematic of Tang active transformers (nMOS). The primary and secondary windings are realized using Wu current reuse active inductors. Coupling is from the interface nodes to the internal nodes of the active inductors.

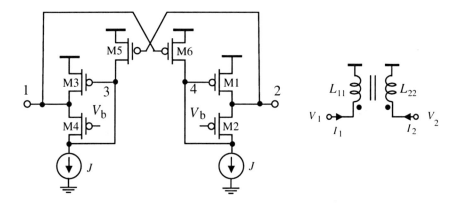

Figure 3.19. Simplified schematic of Tang active transformers (pMOS). The primary and secondary windings are realized using Wu current reuse active inductors. Coupling is from the interface nodes to the internal nodes of the active inductors.

The voltage swing of the active transformer of Fig.3.18 is determined as the followings : To ensure M_6 is in the saturation, $v_{DS,6} \geq v_{GS,6} - V_T$ is required. This is equivalent to $v_1 \leq v_4 + V_T$. Also since $v_1 > V_T$, the voltage swing of the interface nodes of the active transformer of Fig.3.18 is given by $V_T < v_1 \leq v_4 + V_T$. Clearly when $v_1 > v_4 + V_T$, M_6 will enter the triode region. The preceding analysis reveals the drawback of a small input voltage swing when the coupling is from the interface nodes to the internal nodes of active inductors in the construction of gyrator-C active transformers.

Table 3.4. The swing of the input voltage of Tang active transformers (nMOS).

Transformer	Min. voltage	Max. voltage
Fig.3.18	$V_{min} = V_T$	$V_{max} = V_{DD} - 2V_{sat}$
Fig.3.19	$V_{min} = 2V_{sat}$	$V_{max} = V_{DD} - V_T$

The self inductance of the primary winding and the mutual inductance from the primary winding to the secondary winding of a nMOS Tang active transformer implemented in TSMC-0.18μm 1.8V CMOS technology are shown in Fig.3.20 and Fig.3.21, respectively, where the gate voltage of M_1 and M_3 is varied from 1.2 V to 1.8 V with a step 0.1 V. It is seen that both the self and mutual inductances of the active transformer can be tuned by changing V_b with large tuning ranges. The inductance tuning ranges become smaller at high frequencies due to the effect of other parasitic capacitances of the devices.

The active transformers shown in Fig.3.22 and 3.23 are also derived from Wu current-reuse active inductors [104]. The coupling between the active inductors of these transformers is from the internal nodes to the interface nodes of the winding active inductors. It can be shown that the voltage swing of the interface nodes of the active transformers is given by $v_2 > v_3 - V_T$. The upper bound of the voltage swing of the preceding Tang active transformers is removed.

3.3.3 Active Transformers With Low V_{DD} Sensitivity

Similar to CMOS active inductors, the performance of CMOS active transformers is sensitive to supply voltage fluctuation and ground bouncing. In this section, we investigate the sensitivity of the self-inductance of the primary winding of Tang active transformers to the supply voltage.

Consider the nMOS Tang active transformer of Fig.3.24. The supply voltage is varied from 1.7 V to 1.9 V with a step 0.05 V. The self inductance of the primary winding of the active transformer is shown in Fig.3.25. The large sensitivity of the inductance of the transformer to the supply voltage is evident.

To minimize the sensitivity of the inductances of the transformer to its supply voltage, the replica biasing technique proposed in [106] and shown in Fig.3.26 can be employed. The replica-biasing network consists of a sensing circuit made of $M_{9,10,11}$, an auxiliary amplifier, and a reference voltage. M_9 as a V/I

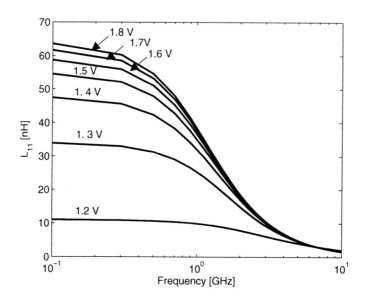

Figure 3.20. Simulated self inductance of the primary winding of nMOS Tang active transformer. The gate voltage of M_1 and M_3 is varied from 1.2 V to 1.8 V with a step 0.1 V (© IEE 2007).

sensor that senses the variation of V_{DD}. $M_{10,11}$ are resistors with resistances $1/g_{m9}$ and $1/g_{m10}$, respectively. They form a voltage divider that provides a proper voltage to the non-inverting terminal of the auxiliary amplifier. An increase in V_{DD} will lead to an increase in v_{SG9}, subsequently the channel current of M_9. The voltage of the non-inverting terminal of the amplifier will also increase. The output of the amplifier will increase the gate voltage of M_9, which ensures that v_{SG9} is kept unchanged ideally. The width of the transistors in the replica-biasing section is usually set to be the same as that of the active inductor section so that they sense the same effect of the variation of V_{DD}.

The quantify the effect of the replica-biasing network analytically, consider the two networks without and with replica biasing shown in Fig.3.27 and Fig.3.28, respectively. The small-signal equivalent circuits of the networks are also shown in the figures. Let the variation of the supply voltage be denoted by v_{dd}. Note $v_{dd} \ll V_{DD}$. To simplify analysis, we neglect the output resistance and all capacitances of the transistors.

Consider first the network without replica biasing. It can be shown that

$$(v_{dd} - v_1)g_m(R_1 + R_2) = v_1. \tag{3.92}$$

Solving for v_1 yields

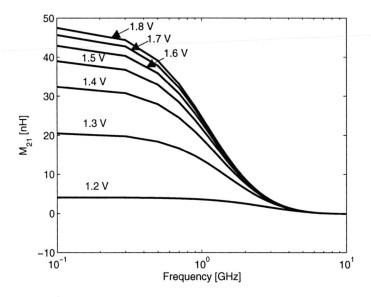

Figure 3.21. Simulated mutual inductance from the primary winding to the secondary winding of nMOS Tang active transformer. The gate voltage of M_1 and M_3 is varied from 1.2 V to 1.8 V with a step 0.1 V (© IEE 2007).

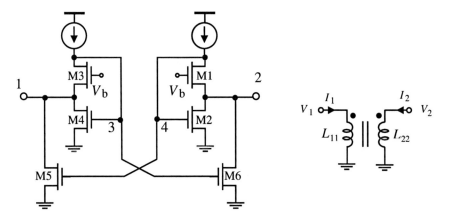

Figure 3.22. Simplified schematic of the variations of Tang active transformers (nMOS). The coupling between the two windings are established from the internal nodes to the interface nodes.

$$v_1 = \frac{g_m(R_1 + R_2)}{1 + g_m(R_1 + R_2)} v_{dd}. \tag{3.93}$$

By assuming that $R_1 = R_2 = \dfrac{1}{g_m}$. Eq.(3.93) becomes

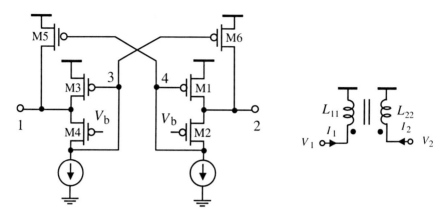

Figure 3.23. Simplified schematic of the variations of Tang active transformers (pMOS). The coupling between the two windings are established from the internal nodes to the interface nodes.

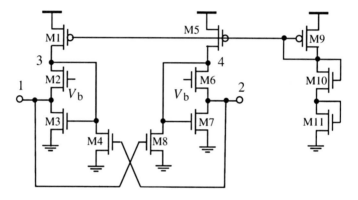

Figure 3.24. Simplified schematic of Tang active transformer (nMOS) without replica biasing.

$$v_1 = \frac{g_m\left(\frac{2}{g_m}\right)}{1 + g_m\left(\frac{2}{g_m}\right)}v_{dd} = \frac{2}{3}v_{dd}. \tag{3.94}$$

Let us now consider the network with replica biasing. Writing KCL at node 2 yields

$$\frac{v_2}{R_2} = g_m(v_{dd} - Av_2). \tag{3.95}$$

Solving for v_2 yields

$$v_2 = \frac{AR_2g_m}{1 + AR_2g_m}v_{dd}. \tag{3.96}$$

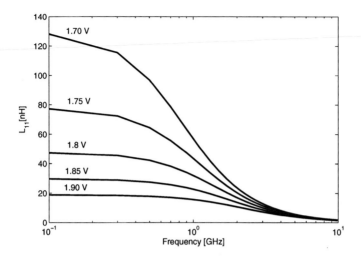

Figure 3.25. Simulated self-inductance of the primary winding of nMOS Tang active transformer. V_{DD} is varied from 1.7 V to 1.9 V with a step 0.05V.

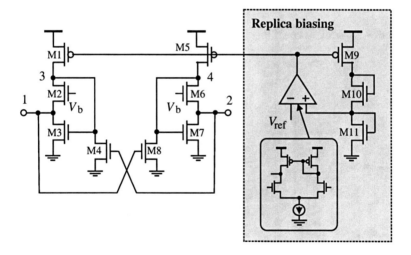

Figure 3.26. Simplified schematic of Tang active transformer (nMOS) with replica biasing to minimize the effect of V_{DD} fluctuation.

Substituting $R_2 = 1/g_m$ into (3.96) yields

$$v_2 = \frac{A}{1+A} v_{dd}$$

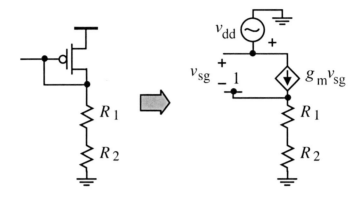

Figure 3.27.   Analysis of replica-biasing - without replica biasing.

$$= \frac{1}{1 + \dfrac{1}{A}} v_{dd}$$

$$= \left[1 + \frac{1}{A} + \frac{1}{A^2} + \ldots \right] v_{dd}. \tag{3.97}$$

It becomes evident from (3.97) that if $A \gg 1$, $v_2 \approx v_{dd}$. As a result,

$$v_{sg} = v_{dd} - v_2 \approx 0. \tag{3.98}$$

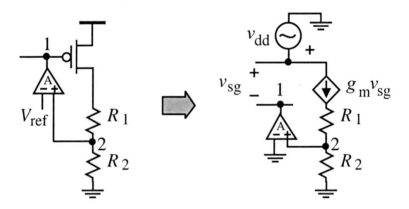

Figure 3.28.   Analysis of replica-biasing - with replica biasing.

The self-inductance of the primary winding L_{11} of the active transformer with replica biasing is shown in Fig.3.29. It is evident that replica-biasing effectively reduces the effect of V_{DD} fluctuation on the performance of the active transformer. Note that the reason why L_{11} without replica biasing differs from that with replica biasing is due to the change of the biasing condition of the gate voltage of M_9.

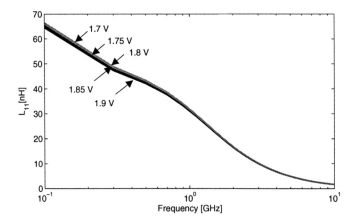

Figure 3.29.   Simulated self-inductance of Tang active transformer with replica biasing. V_{DD} is varied from 1.7 V to 1.9 V with a step 0.05V (© IEE 2007).

3.3.4   Tang Class AB Active Transformers

Class AB active transformers can be constructed by coupling two class AB active inductors using transconductor networks to increase the voltage swing of the transformers. Fig.3.30 shows the simplified schematic of Tang class AB active transformer derived from the class AB active inductors investigated in Chapter 2 [64]. The two class AB active inductors are coupled via parallel coupling transistors M_{9-12}.

Class AB active transformers exhibit a large mutual quality factor at both low and high input currents. This differs from class A active transformers that exhibit a low mutual quality factor at both low and high input currents. This is evident in Fig.3.31.

The average quality factor of the class AB active transformer is obtained by sweeping the current entering the transformer from -10 mA to 10 mA in 50 equal steps, evaluating the instantaneous quality factor using

$$Q(\omega, J) = \frac{\omega}{2} \frac{\partial \phi}{\partial \omega}, \tag{3.99}$$

where ϕ is the phase of the transimpedance from the primary winding to the secondary winding for each current point, and evaluating the expression

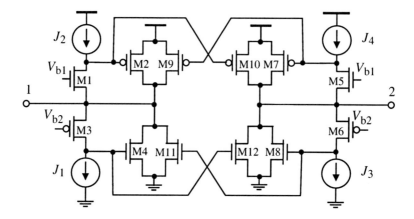

Figure 3.30. Simplified schematic of class AB active transformers

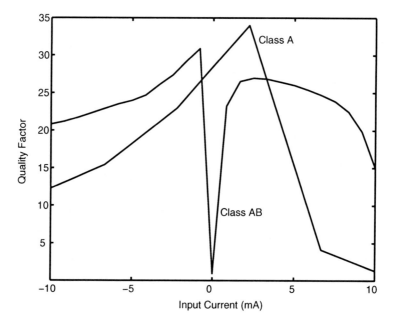

Figure 3.31. Dependence of self and mutual inductances of Tang class A and class AB CMOS active transformers on current swing (© IEEE 2007).

$$\overline{Q(\omega)} \approx \frac{\omega}{2} \left(\frac{1}{15.2 - 0.1} \right) \sum_{j=1}^{50} Q(\omega, J) \Delta J, \qquad (3.100)$$

where $\Delta J = (10 - (-10))/50$ mA. The obtained effective quality factor of the class A transformer is 10.47, while that of the class AB transformer is

15.19. The large average quality factor of class AB transformers ensures that LC VCOs with class AB active transformers will exhibit a low level of phase noise as compared with LC VCOs with class A active transformers, as to be seen in following chapters.

3.4   Chapter summary

A comprehensive treatment of the principles, topologies, characteristics, and implementation of gyrator-C CMOS active transformers has been presented. We have shown that an active transformer can be constructed by coupling two active inductors via a transconductor network. The coupling between two active inductors can be established from the interface nodes to the internal nodes of the active inductors or verse versus. Active transformers constructed by coupling two active inductors from the internal nodes to the interface nodes provide a large voltage swing at the interface nodes. Also, by changing the polarity of the coupling transconductors, active transformers with either positive or negative mutual inductances can be constructed. Lossless gyrator-C active inductors and coupling networks yield lossless transformers while lossy gyrator-C active inductors and coupling networks synthesize lossy transformers.

To provide quantitative measures of the performance of active transformers, a number of critical figure-of-merits have been introduced. These figure-of-merits include stability, frequency range, inductance tunability, turn ratio, coupling factors, voltage transfer characteristics, current transfer characteristics, impedance transformation, quality factors, noise, linearity, stability, supply voltage sensitivity, parameter sensitivity, and power consumption. Frequency range specifies the lower and upper bounds of the frequency between which a transformer characteristic exists. We have shown that the low frequency bound is set by the frequency of the zero of the gyrator-C networks synthesizing the transformers while the upper frequency bound is set by the frequency of the pole of the networks. Both are due to the finite input and output impedances of the transconductors. A key advantage of active transformers over their spiral counterparts is the tunability of their inductances. Both self and mutual inductances of active transformers can be tuned by varying the transconductances of the transconductors. An emphasis has been given to the noise analysis of active transformers as these transformers exhibit a high level of noise power, as compared with their spiral counterparts. Active transformers are active networks that are sensitive to both supply voltage fluctuation and parameter spread. Supply voltage sensitivity and parameter sensitivity have been introduced to quantify the effect of these unwanted variations. The effect of process spread on the performance of active transformers can be compensated for by adjusting the dc biasing conditions of the transformers. Similar to active inductors, the performance of active transformers is dependent of both the dc biasing conditions and the swing of the voltage of the transformers. When active transformers are used in applications such as LC oscillators where signal swing is large, this dependence must be accounted for in design of these oscillators.

The implementation and characteristics of several CMOS active transformers have been investigated in detail. The basic active transformers are evolved from the basic active inductors investigated in Chapter 2. They suffer from a small parasitic parallel resistance and a large parasitic series resistance due to the small output / input impedance of the transconductors of the transformers. This in turn lowers the quality factor of the active transformers. Tang active transformers and their variations are evolved from Wu current reuse active inductors studied in Chapter 2. These active transformers enjoy the advantages of simple configurations and large signal swing. The swing of the input voltage of these active transformers varies with their configurations. To minimize the effect of supply voltage fluctuation on the performance of active transformers, replica-biasing techniques can be employed. We have shown analytically that replica-biasing minimizes the effect of supply voltage fluctuation by maintaining v_{SG} of the biasing transistors constant. This has also been validated using simulation.

PART II

APPLICATIONS OF CMOS ACTIVE INDUCTORS AND TRANSFORMERS

Chapter 4

RF BANDPASS FILTERS WITH ACTIVE INDUCTORS

RF bandpass filters are used extensively in wireless communications. These filters are traditionally implemented using lumped LC, dielectric, and surface acoustic wave (SAW) filters with SAW filters the most widely used, owing to their small size and low power consumption. SAW filters, however, are not compatible with silicon technology. The effort on integrating RF bandpass filters on a silicon substrate is accelerated with the emergence of monolithic spiral inductors and transformers. Two main drawbacks of bandpass filters with spiral inductors and transformers are the low passband center frequency arising from the large spiral-substrate capacitance and the high insertion loss caused by both the skin-effect induced parasitic resistance of the spiral and the substrate eddy-current induced ohmic loss in the spiral at high frequencies. The skin-effect induced loss is proportional to the square root of the frequency of the signal whereas the eddy-current induced loss is proportional to the square of the signal frequency [107]. The former dominates at low GHz frequencies at which most RF bandpass filters operate. Recent RF band select filters are increasingly implemented using Q-enhanced on-chip spiral inductors and transformers to take the advantage of the low-noise and superior linearity of these spiral inductors and transformers and at the same time to minimize the insertion loss [108–112, 107, 23, 113, 24–27]. Active Q-enhancement of spiral inductors is achieved from using a negative resistor that is synthesized using active networks to cancel out the parasitic resistance of the spiral. Because it is difficult to tune the inductance of spiral inductors and transformers in the monolithic implementation of these passive devices, the frequency tuning of these bandpass filters is typically attained using MOS varactors with the downside of a small passband center frequency tuning range. The need for a large silicon area to fabricate spiral inductors and transformers, and the limited capacitance tuning range of MOS varactors greatly increase the cost and limit the robustness of these bandpass filters.

Table 4.1.   CMOS spiral inductor RF bandpass filters.

Ref.	Tech.	Freq. & Tuning [GHz]	1dB comp. [dBm]	IIP$_3$ [dBm]	Power [mW]	Silicon proof
[23] (2002)	0.25μm	2.14, –	-13.4	-4.9	17.5	Y
[24] (2003)	0.35μm	2.1, 13%	-30	-17.5	5.2	Y
[111] (2004)	0.18μm	3.54, 9.44%	-46	-29	130	Y
[26] (2004)	0.35μm	1.035, 11%	-13	–	12.2	Y
[112] (2005)	0.18μm	2.12, –	-3.5	–	143.1	Y
[27] (2006)	0.18μm	2.368, 1.27%	-20	-8.5	8.8	Y

CMOS active inductors have found increasing applications in RF bandpass filters recently, as evident in Table 4.2. These bandpass filters possess a number of attractive characteristics including a high passband center frequency, a large center frequency tuning range, a large and tunable quality factor subsequently a low insertion loss, and a low silicon consumption. The use of active inductors in RF bandpass filters, however, is confronted with several stiff challenges including poor noise performance, a limited dynamic range, and a high level of power consumption.

This chapter examines the design of RF bandpass filters using CMOS active inductors. The chapter starts with a brief presentation of a number of important figure-of-merits that quantify the performance of bandpass filters in Section 4.1. These figure-or-merits include bandwidth, 1-dB compression point, third-order intercept point, noise figure, noise bandwidth, spurious-free dynamic range, frequency selectivity, and passband center frequency tuning range. Section 4.2 deals with the configurations of single-ended and differential active inductor bandpass filters. Section 4.3 examines both the circuit implementation and characteristics of a number of published active inductor bandpass filters. The chapter is summarized in Section 4.4.

4.1   Characterization of Bandpass Filters
4.1.1   Bandwidth

The typical transfer function of second-order bandpass filters is given by

Table 4.2.   CMOS active inductor RF bandpass filters.

Ref.	Freq. [GHz]	IIP$_3$	Tech.	Power [mW]	Silicon proof
Thanachayanont & Payne [114] (1998)	0.9-1.0	-27 dBm	0.8μm	57	N
Yodprasit & Ngarmnil [85] (2000)	1	-25 dBm	0.6μm	–	N
Karsilayan & Schaumann [74] (2000)	1.06	–	0.5μm	4.16	N
Wu *et al.* [45] (2001)	0.9	-15 dBm	0.35μm	12	Y
Thanachayanont & Ngow [65] (2002)	1.0	-9 dBV	0.35μm	2.7	N
Thanachayanont [73] (2002)	2.4-2.6	-10 dBV	0.35μm	2	N
Ngow & Thanachayanont [76] (2002)	1.89-2.09	–	0.35μm	2.0	N
Thanachayanont & Ngow [115] (2003)	2.5	–	0.35μm	2.8	N
Wu *et. al* [36] (2003)	0.4-1.1	-15 dBm	0.35μm	–	Y
Xiao *et al.* [49] (2004)	5.7	270 mV	0.18μm	4.4	Y
Gao *et al.* [51] (2005)	1.6-2.45	-4 dBm	0.25μm	5.1	N
Liang *et al.* [50] (2005)	3.0-4.2	-2.4 dBm	0.18μm	28	Y
Weng & Kuo [62] (2007)	2.0-2.9	–	0.18μm	4	N
Xiao & Schaumann [61] (2007)	3.34-5.72	-1.65 dBm	0.18μm	4.4	Y

$$H(s) = \frac{s\left(\dfrac{\omega_o}{Q}\right)}{s^2 + s\left(\dfrac{\omega_o}{Q}\right) + \omega_o^2}, \tag{4.1}$$

where ω_o is the passband center frequency and Q is the pole quality factor [116]. Note that $|H(j\omega_o)| = 1$ and $|H(j0)| = |H(j\infty)| = 0$. The poles of $H(s)$ are given by

$$p_{1,2} = -\frac{\omega_o}{2Q} \pm j\omega_o\sqrt{1 - \frac{1}{4Q^2}}. \tag{4.2}$$

The two -3dB frequencies of the bandpass filters, denoted by ω_{b1} and ω_{b2} with $\omega_{b2} > \omega_{b1}$, are given by

$$\omega_{b1,b2} = \pm\frac{\omega_o}{2Q} + \omega_o\sqrt{1 + \frac{1}{4Q^2}}. \tag{4.3}$$

The passband width of the second-order bandpass filters, denoted by $\Delta\omega_b = \omega_{b2} - \omega_{b1}$, is shown in Fig.4.1. It follows from (4.3) that

$$\omega_b = \frac{\omega_o}{Q},$$

(4.4)

and

$$\omega_{b1}\omega_{b2} = \omega_o^2.$$

(4.5)

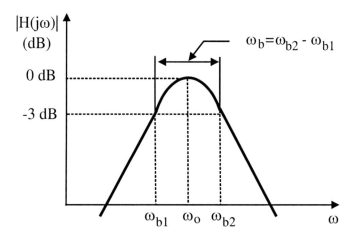

Figure 4.1.   Bodé plot (magnitude) of second-order bandpass filters.

It is evident from (4.4) that the bandwidth of second-order bandpass filters is inversely proportional to the quality factor Q of the filters. The higher the quality factor, the smaller the bandwidth and the better the frequency selectivity.

4.1.2   1-dB Compression Points

The input 1-dB compression point of a filter, denoted by $P_{1dB,in}$, is the power of a single-tone input of the filter at which the output of the filter deviates from the response of a corresponding ideal filter by 1 dB, as illustrated graphically in Fig.4.2. The input 1-dB compression point is a mathematical measure of the harmonic distortion of the filter, which is caused by the drop of the transconductances of the MOS transistors of the filter when the swing of the input of the filter becomes large. The input 1-dB compression point is measured in dBm, that is the ratio of the power of the input signal of the filter to 1 mW in a logarithmic scale.

4.1.3   Third-Order Intercept Points

The input third-order intercept point, denoted by IIP_3, of a filter is the power of a dual-tone input of the filter at which the amplitude of the third-order

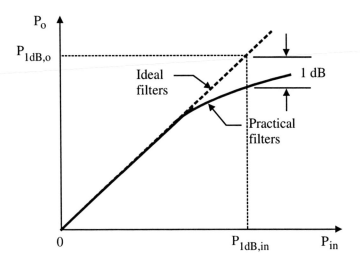

Figure 4.2. 1-dB compression point.

intermodulation component of the output of the filter equals to that of the fundamental component of the output of the filter, as illustrated graphically in Fig.4.3. IIP_3 is also measured in dBm. Note that for practical circuits, although the rate of the increase of the amplitude of the fundamental component of the output is three times that of the third-order intermodulation component of the output, they will never intercept due to the gain compression when P_{in} is high. The third-order intercept point is obtained by extending the plot of the power of the fundamental component of the output and that of the third-order intermodulation component of the output such that they intercept, as shown in Fig.4.3.

4.1.4 Noise Figures

The noise of a filter is typically quantified using its noise figure F defined as the ratio of the signal-to-noise ratio at the input to that at the output of the filter

$$F = \frac{S_i/N_i}{S_o/N_o} = \frac{S_i}{S_o}\frac{N_o}{N_i},$$ (4.6)

where S_i, N_i, and S_o, N_o are the power of the signal and noise at the input and output ports of the filter, respectively. Because

$$N_o = (N_i + N_i')A_p,$$

$$S_o = A_p S_i,$$ (4.7)

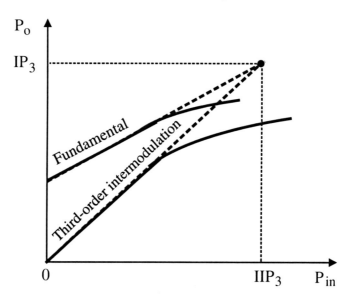

Figure 4.3.   Input third-order intercept point.

where N_i' is the noise power at the input port that is caused by the input-referred noise-voltage and noise-current generators of the filter only, and A_p is the power gain of the filter. Eq.(4.6) becomes

$$F = 1 + \frac{N_i'}{N_i}.\qquad(4.8)$$

If the filter is noiseless, $N_i' = 0$ and $F = 1$ (or 0 dB).

To illustrate the use of (4.8) in the noise figure calculation of filters, consider Fig.4.4 where $\overline{v_n^2}$ and $\overline{i_n^2}$ are the power of the input-referred noise-voltage and noise-current generators of the filter respectively, and R_{in} is the input resistance of the filter. $\overline{v_s^2}$ is the power of the thermal noise of the source resistance R_s.

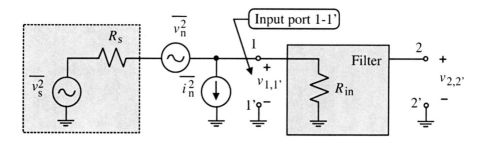

Figure 4.4.   Noise figure calculation of filters.

To compute the noise figure at the input port 1-1' of the filter, we first compute N_i, the noise power at port 1-1' due to $\overline{v_s^2}$ alone.

$$N_i = \left(\frac{R_{in}}{R_{in} + R_s}\right)^2 \overline{v_s^2}. \tag{4.9}$$

To compute N_i', the noise power at port 1-1' due to $\overline{v_n^2}$ and $\overline{i_n^2}$ only, v_s is removed, i.e. v_s is replaced with a short-circuit. Using the principle of superposition, we obtain

$$N_i' = \left[(R_s||R_{in})i_n + \left(\frac{R_{in}}{R_{in} + R_s}\right)v_s\right]^2. \tag{4.10}$$

The noise figure of the filter is obtained by substituting (4.9) and (4.10) into (4.8)

$$F = 1 + \frac{(R_{in}i_n + v_n)^2}{\overline{v_s^2}}. \tag{4.11}$$

Substituting $\overline{v_s^2} = 4kTR_s\Delta f$ into (4.11) and assuming i_n and v_n are uncorrelated, we arrive at

$$F = 1 + \frac{1}{4kT\Delta f}\left(\frac{R_{in}^2}{R_s}\overline{i_n^2} + \frac{1}{R_s}\overline{v_n^2}\right). \tag{4.12}$$

For a perfect impedance matching, i.e. $R_s = R_{in}$, Eq.(4.12) becomes

$$F = 1 + \frac{1}{4kT\Delta f}\left(R_{in}\overline{i_n^2} + \frac{1}{R_{in}}\overline{v_n^2}\right). \tag{4.13}$$

The optimal input resistance of the filter at which the noise figure is minimized can be obtained by differentiating (4.13) with respect to R_{in}

$$\frac{\partial F}{\partial R_{in}} = 0. \tag{4.14}$$

The optimal R_{in}, denoted by R_{in}^*, is obtained from (4.14)

$$R_{in}^* = \sqrt{\frac{\overline{v_n^2}}{\overline{i_n^2}}}. \tag{4.15}$$

4.1.5   Noise Bandwidth

The total noise power at the output of a low-pass filter whose transfer function $H(s)$ is shown in Fig.4.5 is computed from

$$\overline{v_{no,total}^2} = \int_0^\infty |H(j\omega)|^2 S_{in}(\omega)d\omega, \tag{4.16}$$

where $S_{in}(\omega)$ is the single-side power spectral density of the input of the filter at frequency ω. If we use the power spectral density of the output of the filter at $\omega = 0$, denoted by $S_o(0)$, to quantify the output noise power of the filter and assume that the output power spectrum density drops to zero abruptly at frequency ω_n and remains to be zero for $\omega > \omega_n$, as shown in Fig.4.5, the total output noise power of the filter becomes

$$\int_0^\infty |H(j\omega)|^2 S_{in}(\omega)d\omega = S_o(0)\omega_n. \tag{4.17}$$

ω_n is termed the noise bandwidth of the filter. Assume that the input noise is white and $S_{in}(\omega) = 1$ for $0 \leq \omega < \infty$. Further assume that the transfer function of first-order low-pass filters is given by

$$H(s) = \frac{1}{1 + \dfrac{s}{\omega_b}}, \tag{4.18}$$

where ω_b is the -3dB frequency of the filter, as shown in Fig.4.5. Note that $|H(j0)| = 1$. The total noise power at the output of the filter is given by

$$\overline{v_{no,total}^2} = \int_0^\infty \frac{d\omega}{1 + \left(\dfrac{\omega}{\omega_b}\right)^2} = \frac{\pi}{2}\omega_b. \tag{4.19}$$

Making use of the result given in (4.19) in (4.17) and noting that

$$S_o(0) = |H(j0)|^2 S_{in}(0) = 1, \tag{4.20}$$

we obtain the noise bandwidth of first-order low-pass filters [71]

$$\omega_n = \frac{\pi}{2}\omega_b. \tag{4.21}$$

For a second-order bandpass filter whose transfer function is given by

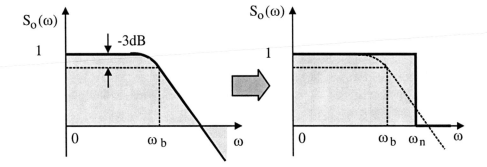

Figure 4.5. Noise bandwidth of first-order low-pass filters.

$$H(s) = \frac{s\left(\dfrac{\omega_o}{Q}\right)}{s^2 + s\left(\dfrac{\omega_o}{Q}\right) + \omega_o^2},$$ (4.22)

we have $|H(j\omega_o)| = 1$, as shown in Fig.4.6. The total output noise power of the bandpass filter whose input is a white noise with a unity power spectral density, i.e. $S_{in}(\omega) = 1$ for $0 \leq \omega < \infty$, is given by

$$\overline{v_{no,total}^2} = \int_0^\infty \frac{\left(\dfrac{\omega_o^2}{Q}\right)^2 d\omega}{(\omega^2 - \omega_o^2)^2 + \left(\dfrac{\omega\omega_o}{\hat{Q}}\right)^2}$$ (4.23)

Making use of the following integration given in [117–119]

$$\int_0^\infty \frac{d\omega}{(\omega^2 - \omega_o^2)^2 + \left(\dfrac{\omega\omega_o}{\hat{Q}}\right)^2} = \frac{\pi}{2}\frac{Q}{\omega_o^3},$$ (4.24)

and noting that $S_o(\omega_o) = |H(j\omega_o)|^2 S_{in}(\omega_o) = 1$, we obtain the noise bandwidth of the 2nd-order bandpass filter

$$\omega_n = \frac{\pi}{2}\frac{\omega_o}{Q}.$$ (4.25)

Recall that the bandwidth of second-order bandpass filters is given by $\omega_b = \dfrac{\omega_o}{Q}$, Eq.(4.25) can thus be written as

$$\omega_n = \frac{\pi}{2}\omega_b. \tag{4.26}$$

The preceding results show that the noise bandwidth of both first and second-order bandpass filters is larger than the bandwidth of respective filters. The relation between ω_n and ω_b of second-order bandpass filters is the same as that of first-order low-pass filters.

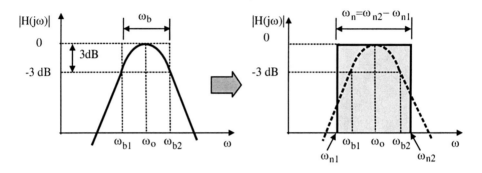

Figure 4.6.   Noise bandwidth of second-order bandpass filters.

4.1.6   Spurious-Free-Dynamic-Range

The sensitivity of a filter is defined as the minimum input signal level at which the filter can yield an acceptable signal-to-noise ratio. The lower end of the spurious-free-dynamic-range is the sensitivity of the filter [17]. The upper bound of the spurious-free-dynamic-range of a filter is the amplitude of a dual-tone input at which the amplitude of the third-order intermodulation terms of the output of the filter do not exceed the noise floor.

4.1.7   Frequency Selectivity and Frequency Tuning

The frequency selectivity of a bandpass filter is directly related to the quality factor of the filter. The higher the quality factor of the bandpass filter, the larger the stopband attenuation and the better the frequency selectivity of the filter.

Frequency tunability is another important design specification of bandpass filters. It is quantified by the range of the variation of the passband center frequency of a bandpass filter in which the quality factor of the bandpass filter meets design specifications. The quality factor and passband center frequency of bandpass filters should be tuned independently.

For a bandpass filter with Q-enhanced spiral inductors, the tuning of the passband center frequency of the filter is typically achieved by varying the control voltage of the varactors of the filter. The quality factor of the bandpass filter is adjusted by varying the dc biasing current of the Q-enhancing negative resistor of the filter. The tuning of these two design quantities is independent of each other. This is one of the intrinsic advantages of spiral inductor bandpass filters.

When a bandpass filter is constructed using active inductors, difficulties arise in achieving the independent tuning of the passband center frequency and quality factor of the filter. As shown in Chapter 2, the inductance of an active inductor is typically tuned by varying the transconductances of the transconductors constituting the active inductor. The quality factor of the active inductor is adjusted by varying the dc biasing current of the compensating negative resistor of the active inductor. Because the variation of the dc biasing current of the negative resistor will usually affect the dc biasing current of the transconductors of the active inductor, the inductance of the active inductor will also change. As a result, the passband center frequency and quality factor of the bandpass filter can not be tuned independently. Active inductor bandpass filters should be designed in such a way that the tuning of the passband center frequency and that of the quality factor of the filters are independent of each other.

4.2   Configuration of Bandpass Filters with Active Inductors

The fact that active inductors are RLC tanks suggests that an active inductor itself is a bandpass filter with the passband center frequency to be the self-resonant frequency of the active inductor. The passband center frequency can be tuned by varying the inductance of the active inductor while the quality factor of the bandpass filter can be adjusted by varying the resistance of the compensating negative resistor of the active inductor. The key advantages of this approach include a simple circuit configuration subsequently a low silicon consumption, a low level of power consumption, a high passband center frequency with a large passband center frequency tuning range, and a large and variable quality factor. The reported active inductor bandpass filters in [74, 46, 36, 115, 51–53, 62, 61] all fall into this category. The basic configuration of these active inductor band-pass filters is shown in Fig.4.7 for single-ended bandpass filters and Fig.4.8 for fully differential bandpass filters.

In Fig.4.7, the input buffer is a transconductor that converts the input voltage of the filter to a current flowing into the downstream active inductor. For RF front-ends, an impedance matching network is also required in front of the input buffer to provide a matching impedance. The active inductor, which is a RLC tank itself, performs the required frequency selection. In order to maximize the quality factor of active inductor bandpass filters, the passband center frequency of the filters should be set to the frequency at which the quality factor of the active inductor of the bandpass filters peaks. A negative resistor is connected in parallel with the active inductor to cancel out the ohmic losses

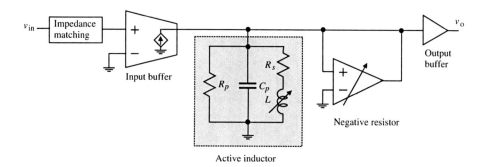

Figure 4.7.   Configuration of single-ended active inductor bandpass filters.

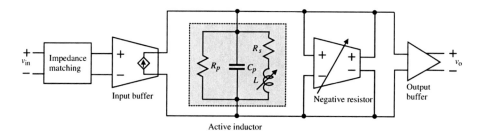

Figure 4.8.   Configuration of differential active inductor bandpass filters.

of the active inductor so as to boost the quality factor of the active inductor and improve its frequency selectivity. The tuning of the center frequency of the bandpass filters is attained by varying the inductance of the active inductor while the quality factor is adjusted by varying the resistance of the compensating negative resistor. The output buffer provides both an adequate driving current and a matching output impedance to the load. The output buffer must also have a large bandwidth so that its impact on the performance of the filter is minimum. Source-follower configurations are typically used in realization of the output buffer due to their low and tunable output impedance and large bandwidth.

4.3   CMOS Active Inductor Bandpass Filters

In this section, we examine the design of CMOS active inductor bandpass filters.

4.3.1   Wu Bandpass Filters

The bandpass filer proposed by Wu *et al.* is based on Wu current-reuse active inductors investigated in Chapter 2. The simplified schematic of the filter is shown in Fig.4.9 [46, 36, 45]. The filter is differentially configured and consists of two Wu current-reuse active inductors whose inductance is tunable by varying J_1. $M_{5,6}$ form a negative resistor to cancel out the parasitic resistances of the active inductor. The resistance of the negative resistor is tunable by varying J_2. M_{7-10} are cascode-configured input transconductors that convert the differential input voltage to a pair of currents flowing into the active inductors. The cascode configuration of these transconductors provides a large output resistance $R_o \approx (g_{m9,10} r_{o9,10}) r_{o7,8}$ such that the loading effect between the transconductor stage and the active inductor stage is minimized. The output of the active inductors is isolated from its load by a source-follower buffer that provides a large bandwidth, an infinite input impedance, and a low output impedance. The output impedance of the output buffer is tunable by varying the dc biasing current of the source follower stage. The center frequency of the bandpass filter is determined from

$$\omega_o \approx \frac{1}{\sqrt{LC_p}}, \tag{4.27}$$

where L is the inductance of the active inductor and C_p is the parasitic parallel capacitance of the active inductor. The center frequency of the bandpass filter can be tuned by varying J_1 and the quality factor of the filter can be adjusted by varying J_2.

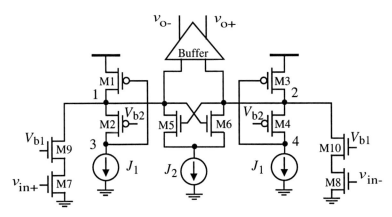

Figure 4.9. Simplified schematic of Wu fully differential active inductor bandpass filter.

A drawback of this bandpass filter is that when J_2 varies (Q-tuning), the currents flowing out of nodes 1 and 2 of the active inductors will also change. As a result, $g_{m1,3}$ will change while $g_{m2,4}$ will remain unchanged. This is echoed with a change in the inductance of the active inductors, subsequently the passband center frequency of the bandpass filter. The tuning of the passband center frequency, however, will not affect the quality of the filter.

4.3.2   Thanachayanont Bandpass Filters

The simplified schematic of the bandpass filer proposed by Thanachayanont is shown in Fig.4.10 [37, 38, 65, 73]. The same bandpass filter was later designed by Gao *et al.* in [51] and extended to high-order bandpass filters in [52, 53]. The filter is differentially configured and consists of two cascode active inductors made of M_{1-6} and $J_{1,2}$. $M_{7,8}$ form a differential negative resistor to cancel out the parasitic resistances of the active inductors. Note that in this configuration, the resistance of the negative resistor is not tunable. The output of the active inductors is buffered with a source-follower configured buffer. $C_{1,2}$ are ac-coupling capacitors whose main function is to block dc signals flowing into the active inductor while allowing ac signals to pass through. The center frequency of the bandpass filter is tuned by varying the inductance of the active inductor, which is achieved by adjusting J_1 and J_2. The results presented in [51–53] show that the interaction between the quality factor and passband center frequency tuning of the bandpass filter is minimum. The quality factor of the bandpass filter remains nearly unchanged when the passband center frequency is tuned.

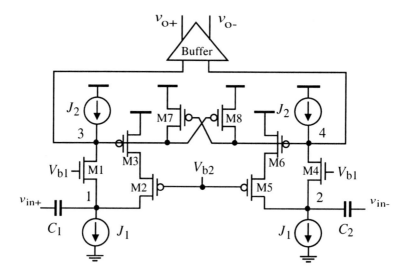

Figure 4.10.   Simplified schematic of 2nd-order Thanachayanont fully differential active inductor bandpass filter.

The preceding Thanachayanont 2nd-order bandpass filter can be employed to construct high-order bandpass filters. As an example, a 4th-order Thanachayanont bandpass filter can be constructed by cascading two Thanachayanont biquads, as shown in Fig.4.11 [65, 52]. 6th-order bandpass filters can also be constructed in a similar way by cascading three Thanachayanont biquads [53].

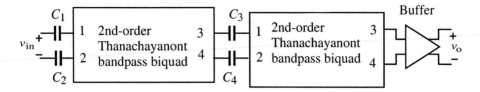

Figure 4.11. 4th-order Thanachayanont fully differential active inductor bandpass filter.

4.3.3 Xiao-Schaumann Bandpass Filters

Karsilayan-Schaumann active inductors investigated in Chapter 2 were used in construction of bandpass filters for RF applications in [75, 48, 49, 61]. The nMOS version of Karsilayan-Schaumann active inductors offer a better frequency response than their pMOS counterparts and were used. The simplified schematic of Xiao-Schaumann bandpass filer is shown in Fig.4.12. M_{1-6} and J_{1-3} form a floating Karsilayan-Schaumann active inductor with its inductance tuned by varying C_I and its quality factor tuned by varying C_Q. It was pointed out in Chapter 2 that one of the key advantages of Karsilayan-Schaumann active inductors is the independent tuning of the inductance and quality factor of the active inductors. $M_{7,8}$ and J_4 form a differentially configured input transconductor that converts a differential input voltage to a differential current flowing into the active inductor. The source-follower configured output buffer drives the load of the bandpass filter and provides a matching impedance, typically 50Ω, to the load.

It was demonstrated in [75, 48, 49, 61] that the bandpass filter implemented in TSMC-0.18μm CMOS technology offered a frequency tuning range from 5.35 GHz to 5.48 GHz. The quality factor tuning range of the filter was from 292 to 665.

4.3.4 Thanachayanont-Payne Bandpass Filters

The bandpass filter proposed by Thanachayanont and Payne in [114] is based on Thanachayanont-Payne cascode active inductors investigated in Chapter 2. The simplified schematic of the bandpass filter is shown in Fig.4.13. M_{1-3} and J_{1-2} form the cascode active inductor. The inductance of the active inductor is controlled by J_1 and J_2. It was shown in Chapter 2 that the parameters of the RLC equivalent circuit of Thanachayanont-Payne cascode active inductors are given by

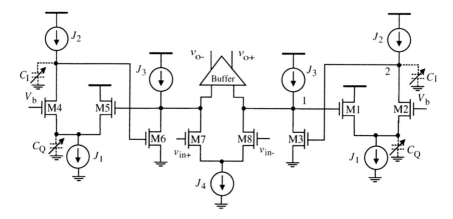

Figure 4.12. Simplified schematic of 2nd-order Xiao-Schaumann fully differential active inductor bandpass filter.

$$R_p = \frac{1}{g_{o2}},$$

$$C_p = C_{gs1},$$

$$R_s = \left(\frac{g_{o1}}{g_{m1}g_{m2}}\right)\frac{1}{g_{m3}r_{o3}}, \tag{4.28}$$

$$L = \frac{C_{gs2}}{g_{m1}g_{m2}}.$$

The self-resonant frequency of the active inductor is given by

$$\omega_o = \frac{1}{\sqrt{LC_p}} = \sqrt{\frac{g_{m1}g_{m2}}{C_{gs1}C_{gs2}}}. \tag{4.29}$$

Since R_p is large, the quality factor of the cascode active inductor is dominated by R_s and can be estimated from

$$Q(\omega_o) \approx \frac{\omega_o L}{R_s} = \frac{\omega C_{gs2}}{g_{o1}}(g_{m3}r_{o3}). \tag{4.30}$$

It is seen from (4.30) that the quality factor of the cascode active inductor is $g_{m3}r_{o3}$ times that of the basic active inductor (without M_3).

Because M_1 and M_3 have the same dc biasing current, a change of J_1 will affect both g_{m1} and g_{m3}. As a result, the tuning of Q by varying g_{m3} via adjusting V_b can not be made independent of the tuning of ω_o. This is a drawback of this bandpass filter.

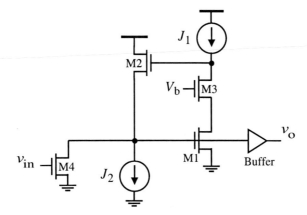

Figure 4.13. Simplified schematic of Thanachayanont-Payne single-ended active inductor bandpass filter.

4.3.5 Weng-Kuo Bandpass Filters

The bandpass filter proposed by Weng and Kuo in [62] is based on Weng-Kuo current-reuse cascode active inductors investigated in Chapter 2. The simplified schematic of the band-pass filter is shown in Fig.4.14. M_{1-3} and J_{1-3} form the current-reuse cascode active inductor. It is seen that g_{m1} is proportional to $J_1 + J_3$ while g_{m3} is only proportional to J_1. Because

$$\omega_o = \sqrt{\frac{g_{m1}g_{m2}}{C_{gs1}C_{gs2}}},$$

$$Q(\omega_o) \approx \frac{\omega_o L}{R_s} = \frac{\omega_o C_{gs2}}{g_{o1}}(g_{m3}r_{o3}),$$

(4.31)

ω_o can be tuned by varying g_{m1} and g_{m2} while Q can be tuned by varying g_{m3} without affecting ω_o. Note that because the tuning of ω_o affects Q, an adjustment of Q will therefore be required after each tuning of ω_o to ensure that the frequency selectivity profile of the filter remains unchanged. The preceding Weng-Kuo bandpass filter can be readily extended to a differential configuration in order to suppress common-mode disturbances.

4.3.6 High-Order Active Inductor Bandpass Filters

The regulated cascode active inductor with a feedback resistor investigated in Chapter 2 was used in construction of a 5th-order bandpass filters by Liang *et al.* in [50]. The simplified schematic of the bandpass filter is shown in Fig.4.15. The grounded inductors are realized using Liang feedback resistance regulated cascode active inductors with resistor feedback shown in Fig.2.54.

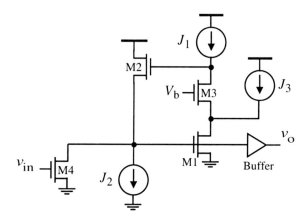

Figure 4.14.   Simplified schematic of Weng-Kuo single-ended active inductor bandpass filter.

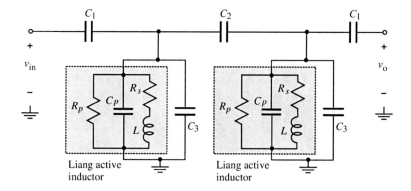

Figure 4.15.   Liang 5th-order active inductor bandpass filter.

It was demonstrated in a UMC-0.18μm implementation of the bandpass filter that the filter had a passband center frequency tuning range of 3.45-3.6 GHz, an insertion loss of 0.2 dB, power consumption of 28 mW, and IIP$_3$ of -2.4 dBm [50]. The noise figure of the bandpass filter varied from approximately 18.5 at 3 GHz to approximately 10 at 4.2 GHz in a nearly linear fashion.

Thanachayanont and Ngow utilized the floating active inductors and class AB active inductors proposed by them (see Chapter 2) to replace the inductors of ladder bandpass filters and constructed inductor-less RF bandpass filters. Shown in Fig.4.16 is a 4th-order bandpass filter [65]. Both class A and Class AB Thanachayanont active inductors investigated in Chapter 2 were used in the implementation of the two floating inductors in Fig.4.16.

In the class A implementation of the bandpass filter in a 0.35μm CMOS technology, the center frequency of the bandpass filter was 2 GHz with the

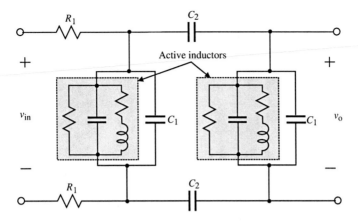

Figure 4.16. Thanachayanont-Ngow active inductor bandpass filter.

quality factor of 25. The spurious-free dynamic range was 52 dB and power consumption was 2 mW.

4.4   Chapter Summary

An in-depth treatment of active inductor bandpass filters has been presented. We have shown that active inductor bandpass filters offer a number of critical advantages over spiral inductor bandpass filters including a large passband center frequency tuning range, a large and tunable quality factor, a low insertion loss, and a low silicon consumption. The applications of these bandpass filters, however, are affected by a number of limitations intrinsic to active inductors including poor noise performance and a limited dynamic range.

The important figure-of-merits that quantify the performance of bandpass filters including bandwidth, 1-dB compression point, third-order intercept point, noise figure, noise bandwidth, spurious-free dynamic range, frequency selectivity, and passband center frequency tuning range have been examined in detail. Using 2nd-order bandpass filters as examples, we have shown that the passband width of 2nd-order bandpass filters is inversely proportional to the quality factor of the filters. The higher the quality factor, the better the frequency selectivity. The relation between the noise bandwidth and the bandwidth of 2nd-order bandpass filters has been derived and is the same as that of first-order low-pass filters.

For active inductor bandpass filters, the tuning of the passband center frequency is carried out by varying the inductance of the active inductors while the frequency selectivity is adjusted by varying the resistance of the Q-enhancing negative resistors of the filters. The tuning of the inductance of active inductors is typically achieved by varying the transconductances of the transconductors constituting the active inductors. The quality factor of the active inductors is tuned by varying the dc biasing currents of the compensating negative resistors of the active inductors. Because the variation of the dc biasing currents of the negative resistors usually affects the dc biasing currents of the transconductors of the active inductors, the tuning of the passband center frequency and that of the quality factor of active inductor bandpass filters often affect each other. Minimizing the interaction between the tuning of the quality factor and that of the passband center frequency, i.e. independent tuning of ω_o and Q of bandpass filters, is essential.

Bandpass filters can be constructed by utilizing the fact that active inductors themselves are RLC tanks. The passband center frequency of these bandpass filters is the self-resonant frequency of the active inductors. The passband center frequency can be tuned by varying the inductance of the active inductors while the quality factor of the bandpass filter can be adjusted by varying the resistance of the compensating negative resistors of active inductors. Bandpass filters constructed in this way offer the attractive advantages of a simple circuit configuration, a low silicon consumption, a large passband center frequency with a large frequency tuning range, a low insertion loss, and a large and tunable quality factor.

The basic configuration of active inductor band-pass filters consists of four blocks - an input transconductor block, an active inductor block, a negative resistor block, and an output buffer block. The input transconductor block converts an input voltage into a current that flows into the active inductor block. The active inductor block performs frequency selection and tuning. This block should be designed in such a way that the quality factor of the active inductor peaks at the center frequency of the filters. The negative resistor block cancels out the ohmic losses of the active inductor block so as to boost the quality factor of the filters and improve frequency selectivity. The output buffer block provides both an adequate driving current and a matching impedance to the load of the filters.

Chapter 5

TRANSCEIVERS WITH ACTIVE INDUCTORS & TRANSFORMERS

This chapter explores the use of CMOS active inductors in transceivers for wireless and wire-line data communications. The chapter starts with an investigation of the use of CMOS active inductors in design of low-noise amplifiers (LNAs) in Section 5.1. We show that due to stringent noise requirements, active inductors are only used at the output node of LNAs where a large capacitance is typically encountered to boost the gain the LNAs. Inductors used for narrow-band impedance matching remains to be spiral inductors. For LNAs in ultra wideband applications, the use of active inductors provides the key advantages of boosting the voltage gain of the amplifiers over a large frequency range. Section 5.2 investigates the use of CMOS active inductors in design of limiting amplifiers encountered in optical front-ends. We show that CMOS active inductors are used extensively in these amplifiers for bandwidth improvement. In Section 5.3, our attention is turned to the design of RF phase shifters where spiral inductors and MMIC passive components are replaced with CMOS active inductors to reduce silicon consumption, improve phase tuning range, and lower insertion loss. Section 5.4 examines the use of CMOS active inductors in wire-line transceivers. Class A and class AB current-mode transmitters, pre-emphasis and post-equalization networks are investigated. Section 5.5 investigates the use of CMOS active inductors in QPSK modulators for Blue-tooth applications. The chapter is concluded in Section 5.6.

5.1 Low-Noise Amplifiers

LNAs are one of the key blocks of RF receivers. The most important design constraints of narrow-band LNAs include (i) impedance matching, (ii) noise figure, and (iii) gain. Spiral inductors are widely used in LNAs for narrow-band impedance matching and gain boosting. Although only a small number of transistors are used in LNAs, the silicon area of LNAs is not small, mainly due to the large area occupied by the spiral inductors for impedance matching and gain boosting. The use of CMOS active inductors in design of LNAs has been explored recently [120–126]. Although the use of active inductors can significantly reduce the silicon consumption of LNAs, the poor noise performance of CMOS active inductors as compared with their spiral counterparts imposes a stiff design challenge. CMOS active inductors have so far only been used at the output node of LNAs where a large capacitance typically exists to boost the gain of the LNAs as the inductance of peaking inductors at this node must be sufficiently large in order to be effective.

An early exploration of the use of CMOS active inductors in design of LNAs was carried out Zhou *et al.* in 1998 where a common-gate configured LNA shown in Fig.5.1 was designed [120]. M_2 is biased in the saturation and provides a dc biasing current for M_1. L_1 and C_{gs1} form a resonator that selects a desirable frequency. Input impedance matching is realized by controlling V_{c2} because $Z_{in} \approx \dfrac{1}{g_{m1}}$. A regulated cascode active inductor is employed at the output node of the LNA and acts as a shunt-peaking inductor to boost the gain of the LNA at frequency set by L_2 and C_o.

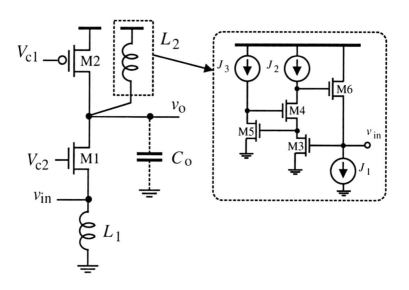

Figure 5.1.   Low-noise amplifier proposed by Zhou *et al.*.

Implemented in a 0.5μm 3.3V CMOS technology, the LNA provided a gain of 20.5 dB when the inductance of the cascode active inductor was set to 50 nH. The noise figure in this case was 3.65 dB and the input impedance was 57 Ω at 1 GHz. The IIP$_3$ was zero and the input 1-dB compression point was -23 dBm.

The LNA proposed by Sharaf and shown in Fig.5.2 is cascode configured [121]. Spiral inductors $L_{1,2}$ are for the impedance matching purpose. V_{c1} provides dc biasing voltage for M_1. Transistor M_2 serves for two purposes : improves the voltage gain and isolates the output and input nodes. Note that without M_2, v_o, which is much larger than v_{in}, would be coupled to v_{in} via C_{gd1}. The cascode active inductor L_3 is employed at the output node of the LNA to improve the gain of the LNA.

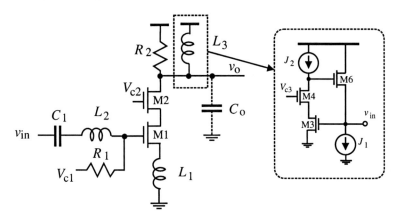

Figure 5.2. Low-noise amplifier proposed by Sharaf.

Implemented in a $0.5\mu m$ 3.3V CMOS technology, the LNA provided a voltage gain of 10.7 dB. The -3 dB bandwidth of the LNA was from 0.66 GHz to 1.16 GHz. The noise figure was between 2.6 dB to 2.9 dB. The IIP$_3$ was -3.8 dBm and the input 1-dB compression point was -33 dBm.

5.2 Optical Front-Ends

In optical communication systems, a transimpedance amplifier (TIA) is typically needed to amplify the photo current of a PIN photo diode, typically $10\mu A$, and converts it to a voltage of approximately 10 mV. This small voltage signal is further amplified to 300 mV approximately by a limiting amplifier (LA), as shown in Fig.5.3, to ensure the reliable operation of the downstream clock and data recovery (CDR) block.

Due to the large bandwidth requirement, inductor peaking with spiral inductors, both shunt peaking and series peaking, is widely used in both transimpedance amplifiers and limiting amplifiers to improve the bandwidth [13, 12]. A main cost of this approach is the high silicon consumption of peaking spiral inductors. Recently, the use of CMOS active inductors for the bandwidth enhancement of limiting amplifiers for optical communications have emerged [41, 40, 79, 80, 127, 82, 128]. The use of active inductors can greatly reduce the silicon cost while satisfying the bandwidth requirement of the amplifiers.

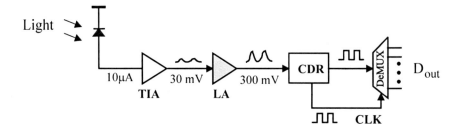

Figure 5.3.   Basic configuration of optical front-ends.

5.2.1   Säckinger-Fischer Limiting Amplifiers

The simplified schematic of the limiting amplifier proposed by Säckinger and Fischer for OC-48 receivers is shown in Fig.5.4 [40, 41]. The amplifier consists of a transimpedance stage, three voltage amplification stages, and one output buffer stage. The common-gate configured transimpedance stage provides a low input impedance of $\dfrac{1}{g_{m1,2}}$ approximately. All amplification stages including the transimpedance stage employ Hara active inductors to boost the bandwidth of the amplifiers. The resistor of Hara active inductors is realized using a pMOS transistor that is powered by a supply voltage V_{DD2} that is larger than V_{DD1} such that the maximum output voltage of $v_{o+,o-}$ can reach V_{DD1}. Note that due to the low load impedance of the differential amplifiers with Hara active inductor loads, the gain of each amplification stage is rather low and is given by

$$A_v \approx \frac{g_{m1,2}}{g_{m3,4}} = \sqrt{\frac{W_{1,2}}{W_{3,4}}} \tag{5.1}$$

at low frequencies where $g_{m1,2}$ and $g_{m3,4}$ are the transconductances of $M_{1,2}$ and $M_{3,4}$ respectively. The output buffer stage is source-follower configured such that it offers both a large bandwidth and a low output impedance of 50Ω to the load. An offset compensation feedback network is required to sense the common-mode voltage at the output of the last voltage amplification stage and to cancel out the input offset current of the transimpedance stage. The compensation network is a low-pass filter with a low cut-off frequency to filter out high-frequency signals at the output of the last amplification stage. Note that only the input offset current of the transimpedance stage is of a critical concern as the input of other amplification stages is much larger.

It was shown in [40, 41] that the limiting amplifier implemented in a 0.25μm CMOS technology offered a bandwidth of 3 GHz and a voltage gain of 32 dB.

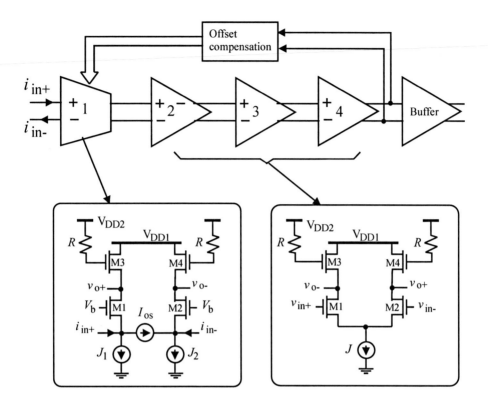

Figure 5.4. Säckinger-Fischer limiting amplifier. $V_{DD2,min} = v_{o+,o-} + V_T$.

5.2.2 Chen-Lu Limiting Amplifiers

The simplified schematic of the limiting amplifier proposed by Chen and Lu for 2.5 Gb/s optical receivers is shown in Fig.5.5 [129, 79]. The input stage and amplification stages of Chen-Lu limiting amplifier are shown in Fig.5.6 and Fig.5.7, respectively. The limiting amplifier consists of six amplification stages and one output buffer stage. All six amplification stages employ Hara active inductors in parallel with their resistor loads for bandwidth enhancement. In addition, regulated cascode is employed in every stage to further boost the bandwidth. To cancel out the offset voltage of the limiting amplifier, a differential RC low-pass filter is employed. The low-pass filter is buffered first so that the large capacitors of the filter will have a minimum loading effect on the last amplification stage. The buffer can also be placed after the low-pass filter, as shown in [128]. The input of the low-pass filter is the output of the last amplification stage whereas the output of the low-pass filter is connected to the offset canceling input terminals of the first amplification of the limiting amplifier.

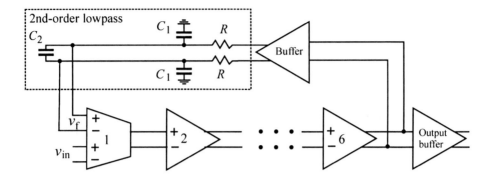

Figure 5.5.   Chen-Lu limiting amplifier. C_2 is an off-chip capacitor due to its large capacitance.

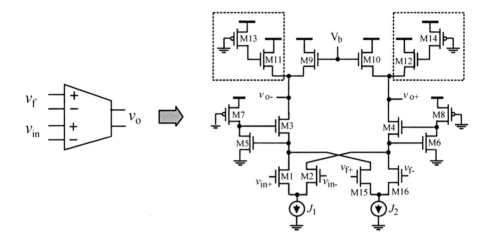

Figure 5.6.   Input stage of Chen-Lu limiting amplifier.

Fabricated in a 0.35μm 3V CMOS technology, the limiting amplifier occupied only 225×500 μm^2 silicon area. The measured bandwidth of the limiting amplifier was 2.2 GHz. The gain of the limiting amplifier was 42 dB.

5.2.3   Wu Limiting Amplifiers

The simplified schematic of the limiting amplifier proposed by Wu *et al.* for 2.5 Gb/s optical receivers is shown in Fig.5.8 [83]. The limiting amplifier consists of one input stage, four amplification stages, one output buffer stage, and a top-detection feedback network for input offset voltage cancellation. All stage employ Wu folded active inductors investigated in Chapter 2 for bandwidth improvement.   To cancel out the offset voltage of the limiting amplifier, a differential RC low-pass filter is employed.

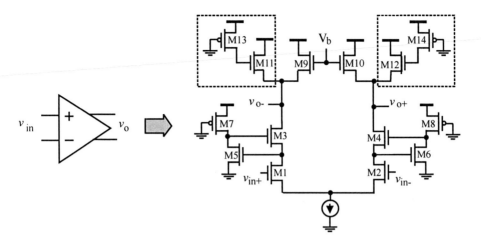

Figure 5.7. Amplification stage of Chen-Lu limiting amplifier.

Fabricated in a 0.25μm 1V CMOS technology, the measured bandwidth of the limiting amplifier was 1.75 GHz. The gain of the limiting amplifier was 40 dB and its power consumption was 4.2 mW.

5.3 Phase Shifters

A phase shifter is a uni-directional network inserted into a signal path such that the phase of the signal at the output of the signal path can be adjusted in a controlled manner [19, 20, 130]. Phase shifters are widely used in RF smart antennas and radar systems for phase adjustment. A well-designed phase shifter possesses the following characteristics : a low insertion loss ($S_{21} \approx 1$), a high return loss ($S_{11} \approx 0$), a large bandwidth so that it has a minimum effect on the bandwidth of the signal path, and a large phase tuning range. A typical configuration of varactor-loaded RF phase shifters is shown in Fig.5.9 [131]. The floating inductors can be implemented using high-impedance metal lines (transmission lines) to minimize the silicon requirement. The tuning of the amount of the phase shift is carried out by varying the control voltage of the shunt varactors. The performance of these phase shifters, however, is affected by the limited phase tuning range due to the small capacitance tuning range of the varactors.

The use of CMOS active inductors in phase shifters emerged recently [20, 102, 95, 132]. In these approaches, CMOS active inductors are used to re-place the passive inductors of phase shifters to take the advantages of CMOS active inductors including a low silicon consumption, a high quality factor subsequently a low insertion loss, and a large inductance tuning range. Also, the location of the floating inductors and grounded varactors are inter-changed

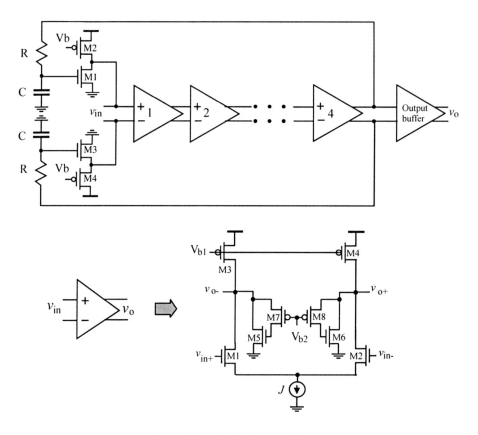

Figure 5.8. Wu limiting amplifier.

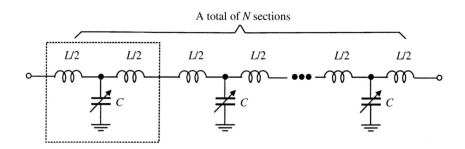

Figure 5.9. Basic configuration of phase shifters.

such that the advantages of the simple configuration of grounded CMOS active
inductors can be fully utilized.

5.3.1 Lu-Liao Active Inductor Phase Shifter

The phase shifter proposed by Lu and Liao in [20] is shown in Fig.5.10. The phase shifter consists of a total of N identical sections, each is made of two floating capacitors and one grounded inductor. The grounded inductors are realized using singe-ended cascode active inductors. Each section provides a phase shift of

$$\Delta\phi = 2\pi f(\sqrt{CL_{max}} - \sqrt{CL_{min}}), \tag{5.2}$$

where L_{max} and L_{min} are the maximum and minimum inductances of the active inductors, respectively, and f is the frequency of the signal passing through the phase shifter. The grounded active inductors are Thanachayanont-Payne cascode active inductor investigated in Chapter 2 whose inductance can be tuned by varying J_1 and J_2. It was shown in [20] that each section of the phase shifter implemented in a 0.18μm CMOS technology provided a maximum phase shift of 60 degrees without a significant reduction of the quality factor of the active inductors. Also, with the Q-enhancement obtained from the cascode configuration, the insertion loss of the phase shifter is mainly due to the series resistance of the on-chip capacitors. The effect of the active inductors on the insertion loss is minimal.

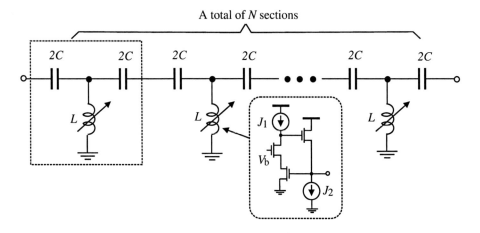

Figure 5.10. Lu-Liao phase shifter using single-ended cascode active inductors.

Measurement results of the phase shifter consisting of eight identical sections with a 4 GHz signal showed that the insertion loss was only 0.3 dB $\pm$ 0.8 dB and the return loss was below -10 dB. The phase shifting range exceeded 360 degrees. The power consumption was between 16 mW and 25 mW, and the silicon consumption of the phase shifter core was only 400$\times$200 μm^2.

5.3.2   Abdalla Active Inductor Phase Shifter

The phase shifter proposed by Abdalla *et al.* in [102, 95, 132] is shown in Fig.5.11. The phase shifter consists of a total of N identical sections, each is made of two floating capacitors, two microstrip lines, and one grounded inductor. The grounded inductors are realized using Abdalla feedback resistance active inductors investigated in Chapter 2.

It was shown in [95, 132] that each section of the phase shifter implemented in a 0.13μm CMOS technology provided a phase shift range from -40 to 34 degrees at 2.5 GHz. The insertion loss is better than 1.1 dB and the return loss is below -19 dB.

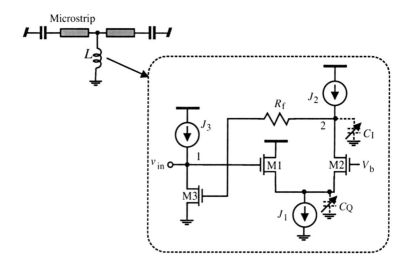

Figure 5.11. Abdalla phase shifter using single-ended cascode active inductors and positive/negative refractive index transmission lines.

5.4   Transceivers for Wire-line Communications

In a point-to-point serial data link, the clock of the link is embedded in the transition of data streams. A serial link transmitter performs the following main tasks : (i) Pre-processing - additional bits are appended to each data byte for embedding both timing information and error checking bits. (ii) Serialization - parallel bits are serialized into an analog waveform. (iii) Pre-emphasis - the serialized analog waveform is pre-distorted in accordance with the characteristics of the channels over which data are transmitted to compensate for the high-frequency loss of the channels. (iv) Transmission - the pre-emphasized output currents are conveyed to channels such that the bit-error-rate of the data received at the far end of the channel meets design specifications.

The speed bottlenecks of a serial link transmitter are its multiplexer where a parallel-to-series conversion takes place and its driver where a large output current or voltage is conveyed to the channels. The former is due to the existence of a N-to-1 multiplexing node at which a large capacitance exists whereas the latter is due to the large signal swing and the large transistors of the driver. To meet timing requirements, inductor-peaking is typically employed at the critical nodes of the transmitter.

5.4.1 Current-Mode Class A Transmitters

Fig.5.12 shows the schematic of the differential class A current-mode driver proposed by Jiang and Yuan in [57]. The driver is preceded by a differential multiplexer that outputs a differential voltage once a multiplexing branch is selected. Diode-connected transistors M_1 and $\overline{M}_1$, and the source connection of transistors M_6 and $\overline{M}_6$ guarantee a low input impedance of $\frac{1}{g_{m1}}||\frac{1}{g_{m6}}$ approximately such that the effect of the large output capacitance of the preceding multiplexer is minimized. Hara active inductors are used at the multiplexing nodes A and $\overline{A}$ to reduce the time constant of the nodes. The low impedance of the channel, typically 50Ω, yields small time constants of the output nodes of the driver despite the large capacitances of the channels. The low impedance seen by the drain of M_2 and $\overline{M}_2$ further eliminates the Miller effect of these devices. In essence, all nodes of the circuit are of a low-impedance characteristic, ensuring that the time constants of the nodes are small. This is one of the key advantages of current-mode circuits.

Consider an input voltage Δv at the input port of the driver. This voltage is represented by $\Delta v/2$ and $-\Delta v/2$ at nodes A and $\overline{A}$, respectively. Assume a perfect device matching between M_j and $\overline{M}_j$, $i = 1, 2, ..., 7$. The corresponding current variation of M_1 and $\overline{M}_1$, denoted by Δi_{D1} and $-\Delta i_{D1}$, respectively, are obtained from

$$\Delta i_{D1} = g_{m1}\left(\frac{\Delta v}{2}\right),$$

$$\Delta i_{\overline{D1}} = -g_{m1}\left(\frac{\Delta v}{2}\right),$$

(5.3)

The current variations of transistors M_2, M_3, and M_5 are given by

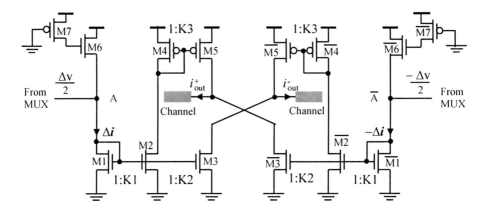

Figure 5.12.   Jiang-Yuan differential current-mode class A transmitter.

$$\Delta i_{D2} = K_1 g_{m1}\left(\frac{\Delta v}{2}\right), \qquad \Delta i_{\overline{D2}} = -K_1 g_{m1}\left(\frac{\Delta v}{2}\right),$$

$$\Delta i_{D3} = K_2 g_{m1}\left(\frac{\Delta v}{2}\right), \qquad \Delta i_{\overline{D3}} = -K_2 g_{m1}\left(\frac{\Delta v}{2}\right), \qquad (5.4)$$

$$\Delta i_{D5} = K_1 K_3 g_{m1}\left(\frac{\Delta v}{2}\right), \quad \Delta i_{\overline{D5}} = -K_1 K_3 g_{m1}\left(\frac{\Delta v}{2}\right),$$

where K_1, K_2, and K_3 are the aspect ratios of the current mirrors $M_{1,2}$, $M_{1,3}$ and $M_{4,5}$, respectively. The output currents are derived from

$$i_{out}^{+} = \Delta i_{D5} - \Delta i_{\overline{D3}} = K_1 K_3 g_{m1}\left(\frac{\Delta v}{2}\right) + K_2 g_{m1}\left(\frac{\Delta v}{2}\right),$$
$$i_{out}^{-} = \Delta i_{\overline{D5}} - \Delta i_{D3} = -K_1 K_3 g_{m1}\left(\frac{\Delta v}{2}\right) - K_2 g_{m1}\left(\frac{\Delta v}{2}\right),$$
$$(5.5)$$

By imposing $K_2 = K_1 K_3$, Eq.(5.5) becomes

$$i_{out}^{+} = K_1 K_3 g_{m1}(\Delta v),$$
$$(5.6)$$
$$i_{out}^{-} = -K_1 K_3 g_{m1}(\Delta v).$$

Eq.(5.6) reveals that i_{out}^{+} and i_{out}^{-} have the same amplitude but opposite polarities.

Fig.5.13 shows the output current of a 10 Gb/s current-mode class A transmitter plotted as a function of the width of the transistor of the Hara active inductor

[57]. The transmitter was implemented in TSMC-0.18μm 1.8V CMOS technology. It is seen that the output current reaches its peak value in less than half the bit time (100 ps).

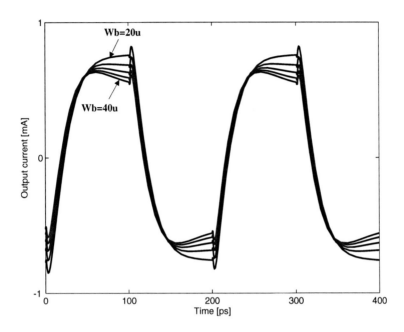

Figure 5.13. Simulated output current of Jiang-Yuan differential current-mode class A transmitter with Hara active inductors (© IEE 2005).

5.4.2 Current-Mode Class AB Transmitters

The drawback of the preceding class A transmitter is its small output current swing, which limits the length of the channels over which data are transmitted. To increase the output current swing, the class AB transmitter shown in Fig.5.14 was proposed by Jiang and Yuan in [133]. It employs a pair of N- and P-differential pairs to achieve a balanced output voltage swing and at the same time to minimize the effect of common-mode disturbances. They are driven by the complementary output voltages of the preceding multiplexer and generate two pairs of differential voltage signals at nodes B and $\bar{B}$, C and $\bar{C}$. These signals are then used to drive the output stage consisting of M_{9-12} that are operated in a push-pull mode. The critical nodes are the gates of M_{9-12}. Hara active inductors are employed at the gates of M_{9-12}. Due to the use of Hara active inductors, v_B and $v_{\bar{B}}$ are shifted down by V_T from V_{DD}. Similarly, v_C and $v_{\bar{C}}$ are leveled up by V_T from the ground, ensuring that transistors M_{9-12} will never turn off. The swing of v_B, $v_{\bar{B}}$, v_C, and $v_{\bar{C}}$ must be properly chosen to ensure that M_{9-12} remain in the saturation such that a large output impedance of the driver exists and the speed penalty of the complete turn-on/off of transistors M_{9-12} is avoided.

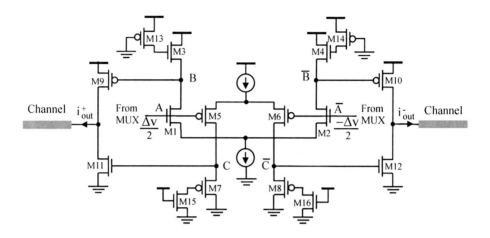

Figure 5.14.   Jiang-Yuan differential current-mode class AB driver.

Consider that the inputs of the driver are given by $v_A = V_{DC} + \dfrac{\Delta v}{2}$ and $\overline{v}_A = V_{DC} - \dfrac{\Delta v}{2}$, where V_{DC} and Δv are the dc and ac components of the input voltage, respectively. Δv is due to the switching of one of the input branches of the preceding multiplexer. Note that the common source of $M_{1,2}$ and the common drain of $M_{5,6}$ are virtual grounds. Assume a perfect device matching between $M_{1,2}$, $M_{3,4}$, $M_{5,6}$, $M_{7,8}$, $M_{9,10}$, and $M_{11,12}$. Because $\Delta i_{D1} = g_{m1}\left(\dfrac{\Delta v}{2}\right)$, $\Delta i_{D2} = -g_{m2}\left(\dfrac{\Delta v}{2}\right)$, $\Delta i_{D5} = -g_{m5}\left(\dfrac{\Delta v}{2}\right)$, and $\Delta i_{D6} = g_{m6}\left(\dfrac{\Delta v}{2}\right)$, we arrive at

$$\Delta i_{D9} = g_{m9}\left[z_{in,n}g_{m1}\left(\frac{\Delta v}{2}\right)\right],$$

$$\Delta i_{D10} = -g_{m10}\left[z_{in,n}g_{m2}\left(\frac{\Delta v}{2}\right)\right],$$

$$\Delta i_{D11} = -g_{m11}\left[z_{in,p}g_{m5}\left(\frac{\Delta v}{2}\right)\right],$$

$$\Delta i_{D12} = g_{m12}\left[z_{in,p}g_{m6}\left(\frac{\Delta v}{2}\right)\right],$$

(5.7)

where $z_{in,p}$ and $z_{in,p}$ denote the impedances of the nMOS and pMOS Hara active inductors, respectively. The ac component of the output current is given by

$$i_{out}^{+} = \Delta i_{D9} - \Delta i_{D11} = (g_{m9}g_{m1}z_{in,n} + g_{m11}g_{m5}z_{in,p})\frac{\Delta v}{2},$$

$$(5.8)$$

$$i_{out}^{-} = \Delta i_{D10} - \Delta i_{D12} = -(g_{m12}g_{m6}z_{in,p} + g_{m10}g_{m2}z_{in,n})\frac{\Delta v}{2}.$$

To have

$$|i_{out}^{+}| = |i_{out}^{-}|,$$

$$(5.9)$$

$$z_{in,n} = z_{in,p}$$

$$(5.10)$$

is required. Eq.(5.8) reveals that the transmitter conveys two currents of the same amplitude but opposite polarities to the channels.

The output current of the transmitter (multiplexer+driver) and the total current drawn by the transmitter from the power supply with input parallel data $D_7D_6...D_1D_0 = 00100011$ are plotted in Fig.5.15. The total current drawn by the transmitter is approximately 10 mA. The swing of the output current exceeds 5 mA.

5.4.3   Pre-Emphasis and Post-Equalization

Pre-emphasis and post-equalization are two effective means to compensate for the high-frequency loss of wire channels. The former reduces the low-frequency components of the signal prior to its transmission whereas the latter amplifies the high-frequency components of the received signal such that an all-pass transfer characteristic of the link is attained. Transmitter pre-emphasis has the advantage of a simple configuration due to the readily availability of the present and previous symbols at transmitters. Because the characteristics of the channels over which data are transmitted are not known a prior, the parameters of the pre-emphasis blocks can not be pre-determined. Tuning is required once the characteristics of the channels are known. Another drawback of pre-emphasis is the high level of power consumption and increased electromagnetic interference due to the large signal swing at the transmitter. The main advantages of post-equalization include a low-level of power consumption due to low signal swing, improved compensation as the effect of the channels on the received data is embedded in the received signal, and a higher speed as the signal swing is much smaller as compared with that at the transmitter side.

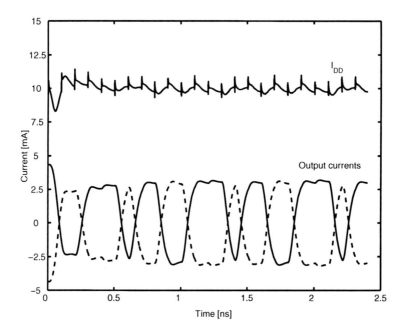

Figure 5.15. Output current and total current of Jiang-Yuan current-mode class AB transmitter.

Pre-Emphasis with Active Inductors

CMOS active inductors have been used extensively in design of pre-emphasis blocks for high-speed serial links and optical links to improve the speed [80, 79, 57, 133]. Transmitter pre-emphasis is often realized using a symbol-spaced finite impulse response filter (FIR) with its characteristics given by

$$V_o(n) = V_i(n) - \sum_{k=1}^{M} a_k V_i(n-k), \quad (5.11)$$

where $V_i(n)$ and $V_i(n-k)$ are the present and past k^{th} symbols, respectively, $V_o(n)$ is the output of the pre-emphasis, a_k is the weighting factor, and M is the number of pre-emphasis taps. The optimal number of pre-emphasis taps is determined by the characteristics of the channels. Constrained by the hardware complexity and the level of the power consumption, the number of taps used in FIR pre-emphasis is usually limited to 3 to 5 [134]. In this section we use the 2.5 Gbps CMOS laser diode driver proposed by Chen *et al.* to demonstrate the use of CMOS active inductors in transmitter pre-emphasis to compensate for the nonlinear effect of laser diodes. Readers can refer to other references such as [135, 57, 133] for the use of CMOS active inductors in pre-emphasis of Gb/s serial links over wire lines.

The single-tap pre-emphasis proposed by Chen *et al.* for 2.5 Gb/s optical links is shown in Fig.5.16 [80]. It employs two wired-OR differential amplifiers, each has a Hara active inductor as their load to boost the bandwidth. The differential amplifier of $M_{1,2}$ represents the current symbol $V(k)$ and that of $M_{3,4}$ represents the last symbol $V(k-1)$. The weighting factor of the pre-emphasis tap is adjusted by varying J_2. Note that the polarity of $V(k-1)$ is arranged such that it is subtracted from $V(k)$. The output current is from the drains of $M_{2,3}$.

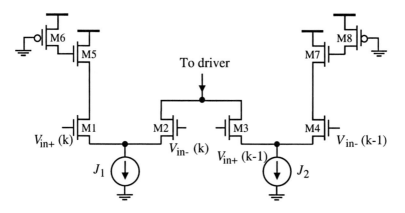

Figure 5.16.   Chen pre-emphasis block with Hara active inductors.

Post-Equalization with Active Inductors

The use of CMOS active inductors in a one-tap post-equalization was used by Chen *et al.* in [54, 55] for 3.125 Gb/s data links. The simplified schematic of the 1-tap equalizer is shown in Fig.5.17 where Wu folded active inductors investigated in Chapter 2 are employed at the output nodes to boost the bandwidth. The equalizer introduces a zero to compensate for the effect of the poles of the channels. The zero is created by the source degeneration formed by the variable capacitors $C_{1,2}$ and the resistor R_1 at the sources of the transistors M_{1-4}.

It was demonstrated in [54, 55] that the equalizer implemented in a 90nm CMOS technology effectively compensated for the high-frequency loss of a 34-inch FR4 backplane, which had a loss of 7 dB at 1.5625 GHz, at data rate 3.125 Gb/s. A significantly increased eye-opening at the receiving end of the backplane was observed.

5.5   Phase Modulators

Quadrature phase-shift-key (QPSK) is the most widely used constant envelope modulation scheme in wireless communications. In QPSK each symbol

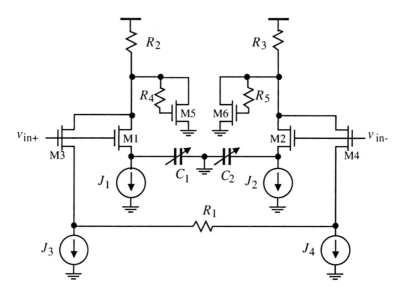

Figure 5.17. Chen equalizer with Wu folded active inductors.

represents two bits. The block diagram of the QPSK modulator proposed by Tang *et. al* is shown in Fig.5.18[63]. An active transformer quadrature oscillator provides the carrier. The multiplexer selects one of four oscillator outputs in accordance with the logic state of the incoming digital stream (2 bits per symbol) [136, 137]. Due to the random nature of the incoming digital bits, the modulation transitions of the carrier may occur at any phase of the carrier, creating sharp transitions [138]. The existence of sharp transactions in the modulated carrier gives rise to the excessive bandwidth of the modulated carrier. In [63], a voltage-controlled delay line (VCDL) was added to select the optimal modulating time instants such that the outputs of the QVCO are sampled at the time instants where the sharpness of the transitions of the sampled data is at the minimum.

The schematic of the current-starving VCDL is shown in Fig.5.19. Two static CMOS inverters are employed with one before and one after the delay stages to improve the waveform. The schematic of the quadrature oscillator employing two Tang active transformers investigated in Chapter 3 is shown in Fig.5.20. The details on the operation of the VCO will be dealt with in Chapter 6.

The modulator was implemented in TSMC-0.18μm 1.8V CMOS technology. A 400 MHz periodic data signal was applied to the input of the modulator. The oscillation frequency of the quadrature oscillator was 1.6 GHz. The phase noise of the quadrature oscillator and that of the modulator are shown in Fig.5.21. It was observed that the phase noise of the active transformer quadrature oscillator

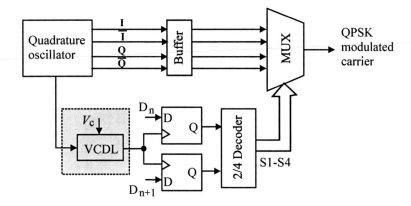

Figure 5.18. Block diagram of QPSK phase modulator.

is approximately -115 dBc/Hz at 1 MHz frequency offset. The phase noise of the modulator is higher than that of the oscillator, mainly due to the noise of the multiplexer.

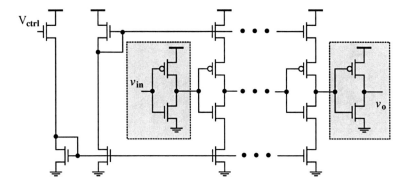

Figure 5.19. Schematic of the voltage-controlled delay line (VCDL).

5.6   Chapter summary

The use of CMOS active inductors in design of various functional blocks of transceivers for wireless and wire-line data communications has been investigated. We have shown that CMOS active inductors have been used in low-noise amplifiers to boost the gain of these amplifiers. Due to the stringent noise requirement of LNAs, CMOS active inductors have only been used at the output node of the LNAs where a large capacitance is typically encountered to boost the gain of the LNAs. Inductors that are used for narrow-band impedance matching at the input of the LNAs remain to be passive to take the advantage of their superior noise performance and linearity.

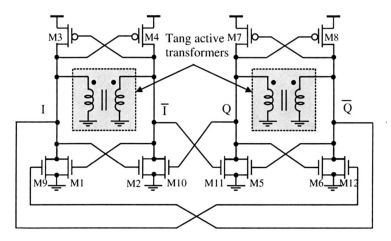

Figure 5.20.   Quadrature oscillator using Tang active transformers.

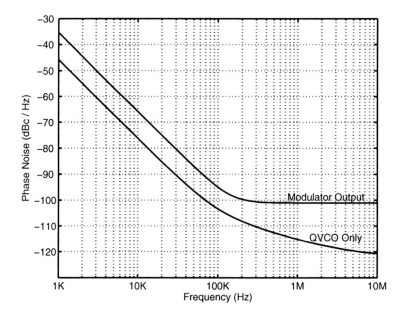

Figure 5.21.   Simulated phase noise of the quadrature LC oscillator with Tang active transformers and that of the modulator (© IEEE 2007).

The use of CMOS active inductors in limiting amplifiers of optical frontends has also been investigated. We have shown that Hara active inductors and Wu folded active inductors are the most widely used active inductors in these amplifiers for bandwidth improvement. Due to the low voltage of differential pairs with Hara active inductor loads at high frequencies, multiple amplification stages are typically required to yield an adequate voltage gain. An offset cancellation is also required to maximize the sensitivity of limiting amplifiers as the input of these amplifiers is small.

The use of CMOS active inductors to replace MMIC passive components in RF phase shifters has also been examined in detail. We have shown that CMOS active inductor-based RF phase shifters offer a number of attractive advantages including a small insertion loss, a large phase tuning range, and a low silicon consumption.

CMOS active inductors have been used in current-mode wire-line transceivers for Gb/s serial links. Because CMOS active inductors in these applications are mainly for propagation delay reduction, they should be placed at the critical nodes of the transceivers where large shunt capacitances exist.

CMOS active transformers have also been used in design of QPSK modulators. In these applications, active transformers are employed to construct quadrature oscillators for the generation of carrier signals to significantly reduce silicon consumption.

Chapter 6

OSCILLATORS WITH ACTIVE INDUCTORS & TRANSFORMERS

This chapter deals with the design of ring and LC tank oscillators using active inductors and transformers investigated in Chapters 2 and 3. The fundamentals of oscillators including voltage or current-controlled oscillators are available in numerous texts, such as [17, 68] and will therefore not be repeated here. Instead, our focus will be given to the circuit implementation of voltage / current-controlled oscillators employing active inductors and transformers for performance enhancement and cost reduction.

The chapter is organized as the followings : Section 6.1 briefly examines the operation of ring oscillators and LC oscillators. The phase noise of oscillators is investigated in detail. Ring oscillators with active inductors are investigated in Section 6.2. Ring oscillators studied in this section include source-coupled ring oscillators, cross-coupled ring oscillators, and Park-Kim ring oscillators. Section 6.3 deals with the design of LC oscillators with active inductors. The use of active transformers in LC oscillators is dealt with in Section 6.4. The design of quadrature LC oscillators using active inductors and transformers is studied in Sections 6.5 and 6.6, respectively. The chapter is concluded in Section 6.8.

6.1 Introduction

The oscillators that are used in data communications over wire channels are mainly ring oscillators due to the need for a large number of clock phases and a large frequency tuning range in these applications. The oscillators that are used in data communications over wireless channels are typically LC oscillators with spiral inductors or transformers, owing to the stringent phase noise requirements and the nature of the narrow-band operation of these systems. The former are usually constructed using a set of voltage-controlled delay cells that are interconnected in a ring format whereas the latter are constructed using cross-coupled amplifiers with LC tank loads.

Oscillators are feedback systems with a positive feedback. The oscillation is initiated by either the noise or disturbances that are either generated internally or coupled externally. Because the amplitude of these noise or disturbances is very small, oscillators can be treated as linear systems at the start of the oscillation where the voltage swing of the oscillators is small. The closed-loop transfer function of the oscillators in this case is given by

$$H_c(j\omega) = \frac{H(j\omega)F(j\omega)}{1 + H(j\omega)F(j\omega)}, \tag{6.1}$$

where $H(j\omega)$, $F(j\omega)$, and $H_c(j\omega)$ are the transfer functions of the forward path, the feedback path, and the closed-loop, respectively. $H_o(j\omega) = H(j\omega)F(j\omega)$ is called the loop gain. Under the condition $H(j\omega)F(j\omega) = -1$, known as Barkhausen criteria, or equivalently

$$\begin{cases} |H(j\omega)F(j\omega)| = 1, \\ \angle H(j\omega)F(j\omega) = -180 \text{ degrees}, \end{cases} \tag{6.2}$$

the system has an infinite gain (oscillation). Although the noise or disturbances generated internally or coupled externally will be amplified with an infinitely large voltage gain theoretically, the finite voltage gain of practical VCOs and the amplitude-limiting mechanism existing in the oscillators ensure that only a finite voltage will appear at the output of the oscillators.

6.1.1   LC Oscillators

An oscillator must have a sufficiently large voltage gain in order to satisfy Barkhausen criteria and start an oscillation in a short period of time. For the LC oscillator shown in Fig.6.1, the voltage gain of a delay stage is given by $A_v \approx -g_m Z$, where Z is the impedance of the LC tank. A large voltage gain exists at the self-resonant frequency of the LC tank. This is because for an ideal LC tank, the impedance of the tank at the resonant frequency ω_o is infinite, i.e. $Z(j\omega_o) = \infty$, where $\omega_o = \dfrac{1}{\sqrt{LC}}$ is the resonant frequency of the LC tank, L and C are the inductance and capacitance of the tank, respectively. For a lossy LC tank, $Z(j\omega_o) = R_p$, where R_p is the shunt resistance of the tank. When neglecting the capacitances of the transistor, the voltage gain of the common-source amplification stage with a lossy LC tank load at the resonant frequency of the LC tank is given by $A_v \approx -g_m R_p$. Although a phase delay of -180 degrees exists ideally, the phase shift is smaller than -180 degrees for practical circuits due to the capacitances of the transistors. Two back-to-back connected common-source amplifiers with LC-tank loads will thus be required to provide a sufficiently large phase shift in order to satisfy Barkhausen criteria.

If the tank is lossy, in order to start an oscillation and sustain the oscillation, a negative resistor connected in parallel with the LC tanks is required to cancel out the resistance of the tanks so that the tanks are lossless. The added negative resistor has two distinct functions : (i) The amplification stages of the LC oscillator have sufficiently large voltage gains to start an oscillation. (ii) The loss of the oscillator is sufficiently small so that an oscillation of a constant amplitude can be sustained.

A widely used differential negative resistor is the cross-coupled transistor pair consisting of $M_{1,2}$ and the tail biasing current source J, as shown in Fig.6.1. A tail biasing current source in this case is needed to provide dc biasing currents for $M_{1,2}$ and to tune the resistance. The resistance of the negative resistor can be tuned by varying the dc biasing current. The resistance of the negative resistor must be at least equal to the resistance of the LC tanks so that the ohmic loss of the tanks can be eliminated completed. For practical circuits, a rule-of-thumb is to set the resistance of the negative resistor three times the resistance of the LC tanks so that the effect of process variation, supply voltage fluctuation, and temperature drift of practical LC oscillators can be accounted for adequately [60, 39].

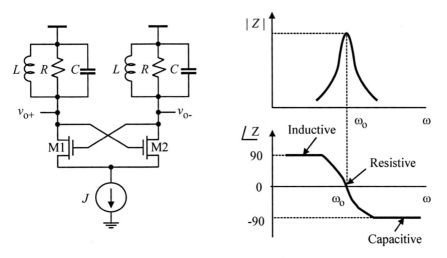

Figure 6.1. Resonance of RLC networks and LC oscillators.

6.1.2 Ring Oscillators

For ring oscillators, such as a static inverter oscillator, a large voltage gain exists when the input of each static inverter stage of the oscillator falls into the transition region of the voltage transfer characteristic curve of the inverter, i.e. $V_{IL} \leq v_{in} \leq V_{IH}$, as shown in Fig.6.2. The voltage gain in this region is given by

$$A_v \approx -(g_{m,n} + g_{m,p})(r_{o,n}||r_{o,p}), \qquad (6.3)$$

where $g_{m,n}$ and $g_{m,p}$ are the transconductances of nMOS and pMOS transistors, respectively, and $r_{o,n}$ and $r_{o,p}$ are the output resistances of nMOS and pMOS transistors, respectively. When the input voltage falls outside the transition region, two cases exist :

- Case 1 - When v_{in} is high, the nMOS transistor is in the saturation and the pMOS transistor is OFF. The voltage gain of the inverter stage drops to $A_v \approx -g_{m,n}r_{o,n}$. To ensure that the nMOS transistor is in the saturation, $v_{DS,n} \geq v_{GS,n} - V_T$ is required. This is equivalent to $v_{in} \leq v_o + V_T$. Clearly, if v_{in} is high enough, the nMOS transistor will be pushed into the triode region. In this case, not only the output resistance of the delay stage will drop from $r_{o,n}$ to $r_{ds,n}$, where $r_{o,n}$ to $r_{ds,n}$ are the channel resistance of the nMOS when the device is in the saturation region and the triode region, respectively, the transconductance of the nMOS transistor will also drop significantly, diminishing the voltage gain.

- Case 2 - When v_{in} is low, the pMOS transistor is in the saturation and the nMOS transistor is OFF. The voltage gain drops to $A_v \approx -g_{m,p}r_{o,p}$. To ensure that the pMOS transistor is in the saturation, $v_{SD,p} \geq v_{SG,p} - V_T$ is required. This is equivalent to $v_{in} \geq v_o + V_T$. If v_{in} is low enough, the pMOS transistor will be pushed into the triode region. Both the output resistance of the delay stage and the transconductance of the pMOS transistor will drop significantly and the voltage gain of the stage will vanish.

Because oscillators are positive feedback systems, an automatic amplitude attenuation and expansion mechanism must be present such that a sustained oscillation with a constant amplitude can be obtained. For the static inverter ring oscillator shown in Fig.6.2, the voltage gain of each inverter stage of the oscillator is very large if the input voltage (noise or disturbances) is small and falls into the transition region of the inverter. The voltage gain of the inverter stages drops drastically when the input voltage of the stages increases and falls outside the transition region. It is this gain drop that limits the amplitude of the output voltages of the inverter stages, giving rise to the peak-to-peak oscillation of the oscillator.

For the LC oscillator shown in Fig.6.1, the voltage gain of the common-source amplifier stages with the LC tank load is limited by the loss of the tank load and the finite transconductance of the transistor. The latter is determined

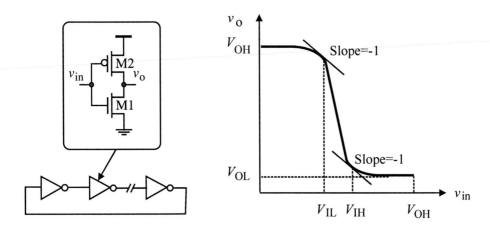

Figure 6.2. Ring oscillators with static inverter delay stages.

by the width of the transistor and its dc biasing current, which is upper-bounded by the tail current of the oscillator. Also, the voltage gain drops drastically when the frequency drifts away from the resonant frequency of the LC tank. Barkhausen criteria are only satisfied at the resonant frequency of the LC tank. This ensures that the oscillation frequency of the LC oscillator will only be the resonant frequency of the LC tank.

The oscillation frequency of a ring oscillator is determined by the number of the delay stages of the oscillator and the average propagation delay of the delay stages of the oscillator.

6.1.3 Phase Noise of Oscillators

The phase noise of oscillators have been studied extensively and various mathematical models of the phase noise of oscillators have been proposed. Perhaps the most widely cited early work on the phase noise of oscillators is the empirical expression by Leeson [139]. It predicts the power of the single-side-band (SSB) phase noise of a LC tank oscillator

$$L(\Delta\omega) = 10\log\left\{\frac{2FkT}{P_s}\left[1+\left(\frac{\omega_o}{2Q\Delta\omega}\right)^2\right]\left(1+\frac{\Delta\omega_{1/f^3}}{\Delta\omega}\right)\right\}, \qquad (6.4)$$

where $\Delta\omega$ is the frequency offset from the oscillation frequency ω_o, Q is the quality factor of the oscillator, F is the excess noise factor, k is Boltzmann constant, T is absolute temperature in degrees Kelvin, P_s is the average power loss of the oscillator, and $\Delta\omega_{1/f^3}$ is the corner frequency between $1/f^2$ and $1/f^3$ regions. As shown in Fig.6.3, $1/f$-noise is up-converted to the vicinity of the oscillation frequency. This region is identified as the $1/f^\beta$-region and affects the phase noise of the oscillators the most. The $1/f^2$-region of the phase noise spectrum is due to the white noise sources of the oscillators whereas the flat region is due to the white noise sources of the output buffers.

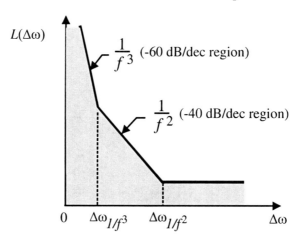

Figure 6.3.   Phase noise spectrum of oscillators.

Weigandt *et al.* showed that an error voltage of an oscillator's output at the nominal time of threshold crossings shifts the actual threshold crossing time by an amount that is proportional to the voltage error and inversely proportional to the slew rate of the output of the oscillator, as shown in Fig.6.4 [140, 141].

$$\overline{\Delta\tau^2} = \frac{\overline{v_n^2}}{(dv/dt)^2}, \tag{6.5}$$

where $\overline{\Delta\tau^2}$ is the timing jitter, $\overline{v_n^2}$ is the power of the noise injected at the threshold-crossing, and dv/dt is the slew rate of the output voltage of the oscillator at the threshold-crossing point.

Razavi showed in [69] that the closed-loop transfer function of oscillators at the frequency that is offset from the oscillation frequency ω_o by $\Delta\omega$ with $\Delta\omega \ll \omega$ is given by

$$\begin{aligned}
|H_c(\omega_o + \Delta\omega)|^2 &\approx \frac{1}{(\Delta\omega)^2 \left|\dfrac{dH_o(j\omega)}{d\omega}\right|^2_{\omega=\omega_o}}, \\
&= \frac{1}{(\Delta\omega)^2 \sqrt{\left[\dfrac{dA(j\omega)}{d\omega}\right]^2_{\omega_o} + \left[\dfrac{d\phi(j\omega)}{d\omega}\right]^2_{\omega_o}}}, \tag{6.6}
\end{aligned}$$

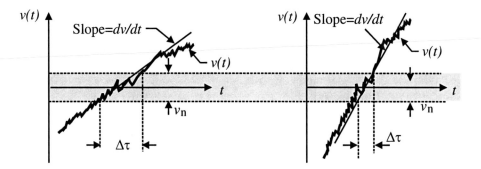

Figure 6.4. Weigandt model of phase noise of oscillators.

where $A(\omega)$ and $\phi(\omega)$ are the magnitude and phase of $H_o(j\omega)$, respectively. By defining the quality factor of the oscillators

$$Q(\omega_o) = \frac{\omega_o}{2} \sqrt{\left[\frac{dA(\omega)}{d\omega}\right]^2_{\omega_o} + \left[\frac{d\phi(\omega)}{d\omega}\right]^2_{\omega_o}}, \tag{6.7}$$

we arrive at

$$|H_c(\omega_o + \Delta\omega)|^2 \approx \frac{1}{4Q^2(\omega_o)}\left(\frac{\omega_o}{\Delta\omega}\right)^2. \tag{6.8}$$

Eq.(6.8) reveals that to minimize the phase noise of oscillators, the quality factor of the oscillators should be maximized.

For saturated ring oscillators, Q is much smaller as compared with that of LC oscillators. This is because at the threshold-crossing of the output voltage of saturated ring oscillators, such as inverter ring oscillators, both the nMOS and pMOS transistors of the inverter stage are in the saturation, exhibiting a large resistance $r_o = r_{o,n}\|r_{o,p}$. Further, there does not exists an Q-enhancement mechanism similar to that in LC oscillators to eliminate the effect of this large resistance. As a result, ring oscillators exhibit a high level of phase noise.

For LC oscillators, because

$$\left.\frac{dA}{d\omega}\right|_{\omega_o} = 0, \tag{6.9}$$

Eq.(6.7) is simplified to

$$Q(\omega_o) = \frac{\omega_o}{2}\left|\frac{d\phi}{d\omega}\right|_{\omega_o}. \tag{6.10}$$

Eq.(6.10) reveals that the quality factor of LC oscillators can be maximized by maximizing the sensitivity of the phase $\phi(\omega)$ with respect to frequency at ω_o.

Because the voltage swing of saturated ring VCOs is usually rail-to-rail and the systems are highly nonlinear, the linearization approach can not characterize the phase noise of saturated ring oscillators accurately. Also, it fails to take into account of the periodically time-varying characteristics of oscillators. As a result, the effect of frequency conversion, especially the up-version of 1/f noise arising from the periodically time-varying characteristics of oscillators, can not be counted for.

Hajimiri and Lee introduced an impulse sensitivity function $\Gamma(\omega_o\tau)$ to quantify the impulse response of oscillators [142]

$$h_\phi(t,\tau) = \frac{\Gamma(\omega_o\tau)}{q_{max}}u(t-\tau), \qquad (6.11)$$

where q_{max} is the maximum charge displacement at the node where the impulse response $h_\phi(t,\tau)$ is measured, $u(t-\tau)$ is the unit step function specifying the time instant at which the noise is injected, τ and t are the noise launch time and response observation time, respectively. Impulse sensitivity function reaches its minima at the peak of the output voltage of the oscillator and its maxima at the threshold-crossing of the output voltage of the oscillator. The periodical operation of oscillators ensures that the impulse sensitivity function is periodic in the observation time t with its period equal to the period of the oscillators. Representing the impulse sensitivity function in its Fourier series expansion

$$\Gamma(\omega_o\tau) = \frac{c_0}{2} + \sum_{m=1}^{\infty} c_m\cos(m\omega_o\tau), \qquad (6.12)$$

we arrive at the phase noise induced by a noise current $i_n(t)$

$$
\begin{aligned}
\phi(t) &= \int_{-\infty}^{\infty} h_\phi(t,\tau)i_n(\tau)d\tau \\
&= \frac{1}{q_{max}} \int_{-\infty}^{t} \Gamma(\omega_o\tau)i_n(\tau)d\tau \\
&= \frac{1}{q_{max}} \left[\frac{c_0}{2} \int_{-\infty}^{t} i_n(\tau)d\tau + \sum_{m=1}^{\infty} c_m \int_{-\infty}^{t} \cos(m\omega_o\tau)i_n(\tau)d\tau \right].
\end{aligned}
$$
$$(6.13)$$

Eq.(6.13) provides a theoretical foundation for analysis of the up-conversion of low-frequency noise.

For a white noise source with its PSD given by

$$S(\omega) = \frac{\overline{i_n^2}}{\Delta f}, \tag{6.14}$$

the SSB computed from (6.13) is given by

$$L(\Delta\omega) = \frac{\Gamma_{rms}^2}{q_{max}^2} \frac{\overline{i_n^2}/\Delta f}{2(\Delta\omega)^2}. \tag{6.15}$$

For a $1/f$ noise source with its PSD given by

$$S(\omega) = \overline{i_{n,1/f}^2} = \overline{i_n^2} \frac{\omega_{1/f}}{\Delta\omega}, \tag{6.16}$$

where $\omega_{1/f}$ is the corner frequency of the $1/f$ noise, the SSB is given by

$$L(\Delta\omega) = \frac{c_0^2}{q_{max}^2} \frac{\overline{i_n^2}/\Delta f}{4(\Delta\omega)^3} \omega_{1/f}. \tag{6.17}$$

6.2 Ring Oscillators With Active Inductors

Voltage-controlled ring oscillators offer many attractive advantages over their LC counterparts including full compatibility with standard CMOS processes, a large frequency tuning range, a large number of complementary phases, a low level of power consumption, and a low silicon consumption. The stringent timing jitter constraint of high-speed data communications require that these ring oscillators be configured in a fully differential way such that the common-mode noise generated internally by the oscillators or coupled externally to the oscillators is suppressed effectively.

Eq.(6.5) reveals that an increase in the slew-rate at the threshold-crossing of the output voltage of VCOs will lower the timing jitter, as depicted graphically in Fig.6.4. The output voltage of ring VCOs should therefore be designed to have fast rising and falling edges such that the transition window during which circuit noise contributes the most to the timing jitter of the oscillators is minimized.

It is well understood that the slope of the rising and falling edges of the output voltage of a conventional ring VCO is set by the RC time constant of the output node of the VCO. Although this RC time constant can be reduced by varying the width of the transistors of the delay cell of the VCO, the improvement is rather moderate. This is because an increase in the width of the transistors of the delay cell of the VCO, though lowering R, increases C at the same time as well. As a result, the net reduction in the time constant of the output node of the delay stage of the VCO is rather marginal.

This section looks into the use of active inductors in design of ring oscillators to improve the slope of the output voltage of the oscillators at threshold crossings.

6.2.1 Source-Coupled Ring VCOs

The schematic of the delay cell of source-coupled differential ring VCOs is shown in Fig.6.5. Common-mode disturbances either generated internally or coupled externally, such as supply voltage fluctuations, are suppressed effectively by the differential configuration of the delay cell. The effect of ground bouncing is also minimized as the delay cell is isolated from the ground rail by the biasing tail current source. The impedance of the active loads is voltage-dependent and varies with the swing of the output voltage v_o. When v_o is low, the load transistors are in the saturation and behave as current sources, exhibiting a large impedance. When v_o is high, the load transistors enter the triode region and behave as resistors, exhibiting a low impedance. The time delay of the delay cell can be varied in three different ways :

- Bias the active load transistors $M_{3,4}$ in the triode region by keeping V_{c2} low and control their resistance by varying the gate voltage V_{c2}. Because $M_{3,4}$ behaves as voltage-controlled resistors only when they are in the triode, the swing of the output voltage of the delay cell must be kept small. This can be achieved by keeping the number of the delay stages small such that $v_{SD3,4} < V_{sat}$. A key advantage of this approach is that the total current drawn by the delay cell from the power supply is constant and is set by the tail current source. As a result, the switching noise generated by the delay cell is minimized.

- Bias the active load transistors $M_{3,4}$ in the saturation and control their channel currents by varying the gate voltage V_{c2}. Similar to the case in which $M_{3,4}$ are in the triode, the total current drawn by the delay cell is constant and is set by the tail current source. The switching noise generated by the delay cell is minimized. It should be noted that the preceding two approaches are often used without a clear boundary as the mode of the operation of $M_{3,4}$ is dependent of both the biasing voltage V_{c2} and the output voltage. Also, the preceding two methods vary the propagation delay of the delay cell by only changing the charge time of the output nodes. The discharge time is fixed.

- Vary the gate voltage V_{c1} of the tail transistor M_5. Because $g_{m1,2}$ vary with V_{c1} in this case, the delay of the delay stage experiences a large variation, echoed with a large frequency tuning range. A downside of this approach is that when the VCO is used in a phase-locked loop with a bang-bang phase detector, the control voltage fluctuates in the vicinity of its nominal value.

The total current of the delay cell injected into the ground rail and drawn from the supply voltage is no longer constant. As a result, the switching noise generated by the delay cell increases. Note that the propagation delay of the delay cell is varied by changing the discharge time in this case.

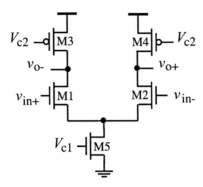

Figure 6.5. Schematic of source-coupled ring VCO delay cell with active loads.

The delay of the preceding source-coupled delay cell is set by the RC time constant of the output nodes. To further reduce the delay, active inductors can be used to replace the active loads of the delay cell, as shown in Fig.6.6 where Hara active inductors are used as the loads of the delay cell [41, 40]. The time constant of the output nodes of the delay cell is now determined by the RLC networks at the output nodes. In addition to the time constant reduction obtained from the inductive peaking, the added Hara active inductors further improve the performance of the delay cell in the following aspects :

- Although Hara active inductors employ two transistors with the pMOS transistor biased in the triode and behaving as a voltage-controlled resistor, only the nMOS transistor of the active inductors is connected to the output nodes of the delay cell. As a result, the output capacitance of the delay cell is approximately the same as that of the delay cell with the active loads.

- The low impedance looking into Hara active inductors given by $\dfrac{1}{g_{m5,6}}$ lowers the time constant of the output nodes of the delay cell from $\tau_o \approx C_o\left(r_{o1,2}||r_{o3,4}\right)$ without the active inductors to $\tau_o \approx \dfrac{C_o}{g_{m5,6}}$ with the active inductors, where C_o is the capacitance encountered at the output node. A nMOS-latch formed by $M_{7,8}$ and shown in Fig.6.6 can also be employed to further improve the performance of the delay cell [81].

The frequency tuning of source-coupled ring VCOs with Hara active inductor loads can be carried out in two ways :

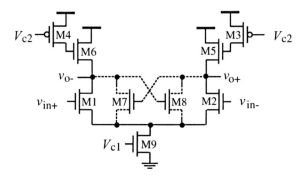

Figure 6.6.   Schematic of source-coupled ring VCO delay with Hara active inductor loads.

- Approach 1 - Vary the gate voltage of $M_{3,4}$. Because $M_{3,4}$ are in the triode and behave as voltage-controlled resistors, a change of V_{c2} will change the resistance of the resistor of Hara active inductors. It was shown in Chapter 2 that the inductance of Hara active inductors can be tuned by varying its resistor R with a small inductance tuning range. The small inductance tuning range of Hara active inductors results in a small frequency tuning range of the VCOs obtained from varying R, as demonstrated in [143].

- Approach 2 - Vary the gate voltage of the tail biasing transistor M_9. A large frequency tuning range can be obtained in this way. As pointed out earlier that a downside of this approach is that the total current by the delay cell injected to the ground rail and drawn from the power supply is no longer constant. This is echoed with an increase in the switching noise generated by the delay cell.

   The delay cell with Hara active inductor loads suffers from a voltage swing loss of at least V_T because when v_o exceeds $V_{DD} - V_T$, the nMOS transistor of Hara active inductors switches off, disabling the active inductors. This drawback can be eliminated by adding a pull-up pMOS latch, as shown in Fig.6.7. The added pMOS latch will continue to charge the output nodes even when Hara active inductors switch off.

6.2.2   Cross-Coupled Ring VCOs

   A main drawback of the preceding source-coupled delay cells is the need for a biasing tail current source. As pointed out in [142] that the flicker noise of the tail current source of the delay stage of a ring oscillator will be up-converted to the vicinity of the oscillation frequency of the ring oscillator, deteriorating the phase noise of the oscillator. Also, the output voltage swing of the oscillator is limited by the voltage drop across the tail current source.

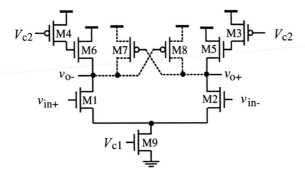

Figure 6.7. Schematic of source-coupled ring VCO delay with Hara active inductor loads and a pull-up pMOS latch.

The cross-coupled delay cell shown in Fig.6.8 removes the need for a tail biasing current source [144, 145]. It operates in a full rail-to-rail swing mode. A nMOS-latch is often employed to speed up the transition of the output voltages, reducing the transient duration in which device and supply voltage noise contributes the most to the timing jitter of the oscillators. The latch also ensures that the output voltage is less sensitive to both supply voltage variation and ground bouncing once the latch is established.

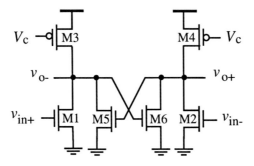

Figure 6.8. Cross-coupled ring VCO delay cell with active loads and a pull-down nMOS latch.

$M_{3,4}$ can be either in the triode or in the saturation, depending upon V_c. Also, their region of operation is dependent of the output voltage swing of the delay cell. Frequency tuning is achieved by varying V_c that controls $r_{ds3,4}$ when $M_{3,4}$ are in the triode or $i_{DS3,4}$ when $M_{3,4}$ are in the saturation. A drawback of cross-coupled ring VCOs is waveform asymmetry. As shown in Fig.6.9, the rise time of the cross-coupled VCOs varies with the control voltage while the fall time remains unchanged. As a result, the duty cycle of the waveform varies with the control voltage. Another drawback is the timing-varying current injected into the ground rail, increasing the level of switching noise.

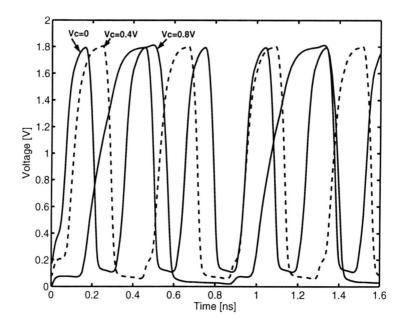

Figure 6.9. Simulated output voltage of a 4-stage cross-coupled ring VCO with active loads and a nMOS pull-down latch.

To reduce the time delay of the cross-coupled delay cells, the active loads can be replaced with Hara active inductor loads, as shown in Fig.6.10 [81]. The added active inductors form RLC networks at the output nodes of the delay cell. Inductor shunt peaking lowers the time for charging and discharging the output nodes. The tuning of the oscillation frequency is achieved by varying the value of the resistor of Hara active inductors via adjusting V_c. This delay cell, however, suffers from two main drawbacks : (i) The frequency tuning range is small as the resistance of the resistor of Hara active inductors has a marginal effect on the inductance of the active inductors. (ii) The voltage headroom loss of at least V_T at the output nodes of the delay cell because v_b is upper-bounded by $V_{DD} - V_T$.

The preceding voltage headroom loss can be eliminated by connecting a pull-up pMOS latch in parallel with the active inductor load, as shown in Fig.6.11 [59]. The pull-up pMOS latch takes over the charging process of the output node when the output voltage exceeds $V_{DD} - V_T$ at which the nMOS transistor of Hara active inductors enters its cut-off mode. Because no change is made to the frequency tuning mechanism, the drawback of the small frequency tuning range remains.

Fig.6.12 and Fig.6.13 compare the waveform of the output voltage of two 4-stage cross-coupled ring VCOs, one with resistor loads and the other with

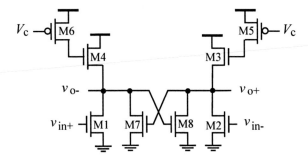

Figure 6.10. Schematic of cross-coupled ring VCO delay cell with Hara active inductor loads and a pull-down nMOS latch.

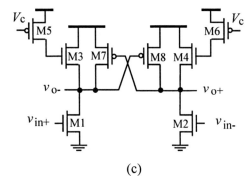

(c)

Figure 6.11. Schematic of cross-coupled ring VCO delay cell with Hara active inductor loads and a pull-up pMOS latch.

Hara active inductor loads. The corresponding transistors of two ring VCOs were chosen to have the same dimension so that a fair comparison was obtained. Both VCOs were implemented in TSMC-0.18μm 1.8V CMOS technology and analyzed using SpectreRF with BSIM3V3 device models. It is seen that the oscillation frequency of the VCO with Hara active inductor loads is approximately twice that of the VCO with the resistor loads. The loss of the swing of the output voltage of the VCO with Hara active inductor loads is also evident. Fig.6.14 plots the output voltage of a 4-stage cross-coupled ring VCO with Hara active inductor loads and the pull-up pMOS-latch. It is observed that the swing of the output voltage of the VCO is rail-to-rail. Also, the added pMOS-latch improves the slew rate of the output voltage, reflected by the increased time duration of the Logic-0 and Logic-1 stages of the output voltage.

6.2.3 Park-Kim Ring VCOs

The delay cell proposed by Park and Kim in [146] and shown in Fig.6.15 controls the propagation delay of the delay cell by varying the intensity of

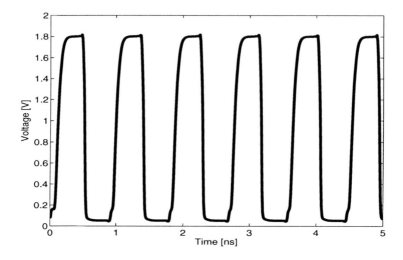

Figure 6.12.   Simulated output voltage of a 4-stage cross-coupled ring VCO with resistor loads.

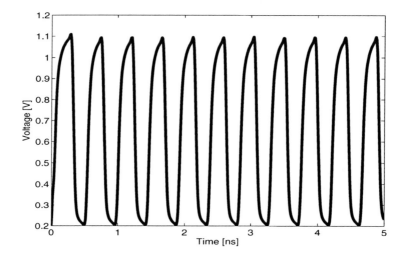

Figure 6.13.   Simulated output voltage of a 4-stage cross-coupled ring VCO with Hara active inductor loads.

the latch. An advantage of doing so is that both the rise and fall times are controlled by V_c. As a result, an improved waveform symmetry over a large range of the control voltage is obtained, as evident in Fig.6.16. This is an important advantage of Park-Kim ring VCOs over the cross-coupled ring VCOs investigated earlier.

In [58], a pair of Hara active inductors were connected in parallel with the pull-up pMOS transistors of Park-Kim ring VCO delay cell to provide an

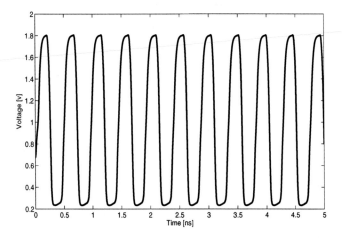

Figure 6.14. Simulated output voltage of a 4-stage cross-coupled ring VCO with Hara active inductor loads and a pull-up pMOS latch.

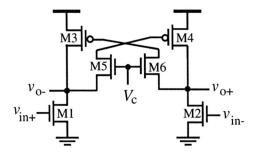

Figure 6.15. Schematic of single-loop Park-Kim ring VCO delay cell.

additional charge path to compensate for the slow turn-on of the pull-up pMOS transistors and to provide active inductors at the output nodes of the delay cell, as shown in Fig.6.17. The frequency tuning mechanism of the delay cell is the same as that of Park-Kim ring VCO delay cell so that the waveform symmetry property of Park-Kim ring VCO delay cell is preserved.

Figs.6.18 and 6.19 plot the simulated output voltage of two 8-stage Park-Kim ring VCOs, one with Hara active inductor loads and the other without. The VCOs were implemented in TSMC-0.18μm 1.8V CMOS technology and analyzed using SpectreRF with BSIM3V3 device models. The size of corresponding transistors in the delay cells was made identical in order to obtain a fair comparison. The improvement in the oscillation frequency from employing Hara active inductors is evident. The oscillation frequency of the VCO with

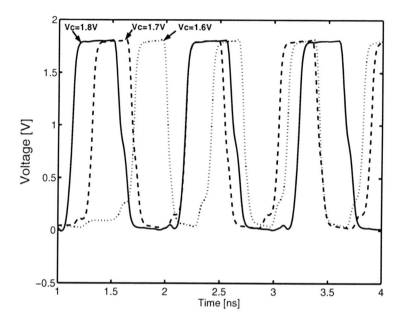

Figure 6.16.   Simulated output voltage of a 8-stage single-loop Park-Kim ring VCO.

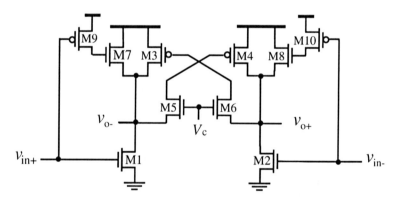

Figure 6.17.   Schematic of single-loop Park-Kim ring VCO delay cell with Hara active inductor loads.

Hara active inductor loads is approximately 2.5 times that of the VCO without active inductor loads.

The oscillation frequency of Park-Kim VCOs can be increased by employing a dual-loop configuration, as shown in Figs.6.20 and 6.21 [146, 147]. The additional pair of inputs are driven by the inputs of the preceding stage to provide a pair of pre-driving signals that arrive earlier than the main inputs of

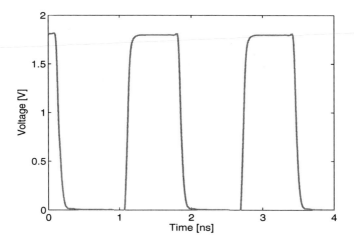

Figure 6.18. Simulated output voltage of a 8-stage single-loop Park-Kim ring VCO.

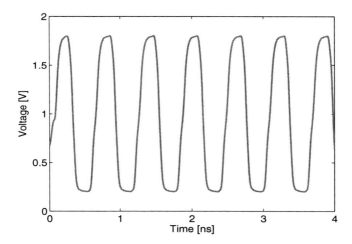

Figure 6.19. Simulated output voltage of a 8-stage single-loop Park-Kim ring VCO with Hara active inductor loads.

the delay stage so that the slow turn-on of the pMOS transistors of the delay cell can be compensated for effectively.

The delay of the dual-loop Park-Kim VCO delay cell can be further increased by replacing the pull-up transistors $M_{7,8}$ in Fig.6.21 with a pair of Hara active inductors, as shown in Fig.6.22. The dual-loop Park-Kim delay cell with Hara active inductor loads bears a strong resemblance to the dual-loop Park-Kim delay cell in the following aspects :

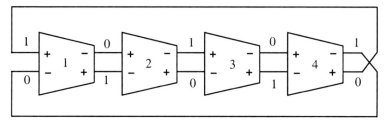

(a) Single-loop ring VCOs

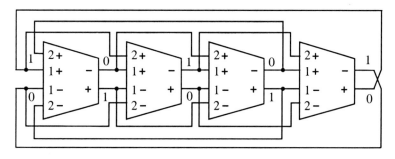

(b) Dual-loop ring VCOs

Figure 6.20.   Configuration of single-loop and dual-loop ring VCOs.

■ Frequency tuning mechanism - Both delay cells adjust the time delay by varying the intensity of the latch. This not only enables the secondary inputs of the delay cells to be used to compensate for the slow turn-on of the pull-up pMOS load and ensures that the frequency tuning range of Park-Kim ring VCOs with Hara active inductor loads is comparable to that of corresponding Park-Kim ring VCOs, it also preserves the waveform symmetry property of Park-Kim ring VCOs.

■ Output time constant - The output capacitance of the dual-loop Park-Kim delay cell with Hara active inductor loads is approximately the same as that of the dual-loop Park-Kim delay cell as the number of the transistors at the output node is the same.

   The output resistance of the dual-loop Park-Kim delay cell with Hara active inductor loads given by $\dfrac{1}{g_{m7,8}}$, however, is smaller. As a result, a smaller time constant at the output nodes of the delay cell is obtained.

   Fig.6.23 and Fig.6.24 compare the waveforms of the output voltage of two 8-stage dual-loop Park-Kim VCOs with and without Hara active inductor loads. The delay cells of both VCOs have the same transistor size and biasing condi-

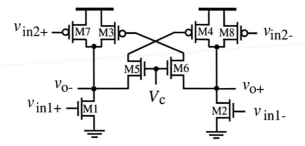

Figure 6.21. Schematic of dual-loop Park-Kim ring VCO delay cell.

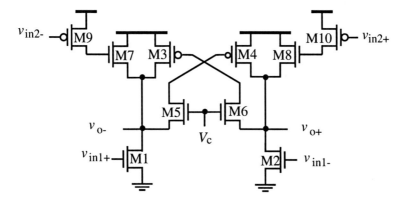

Figure 6.22. Schematic of dual-loop Park-Kim ring VCO delay cell with Hara active inductor loads.

tions. They were implemented in TSMC-0.18μm 1.8V CMOS technology and analyzed using SpectreRF with BSIM3V3 device models. The improvement in the oscillation frequency obtained from using Hara active inductor loads is evident.

6.3 LC Oscillators With Active Inductors

As compared with ring oscillators, LC oscillators with spiral inductors offer a key advantage of a low level of phase noise. They are widely used in wireless communications where a stringent constraint on phase noise exists. Two widely used configurations of LC oscillators with spiral inductors are shown in Figs.6.25 and 6.26 [148, 6]. Both are differentially configured in order to suppress common-mode disturbances effectively. The LC oscillator in Fig.6.26 employs two negative resistors implemented using a cross-coupled nMOS pair ($M_{1,2}$) and a cross-coupled pMOS pair ($M_{3,4}$) to compensate for the loss of the tanks. The oscillation amplitude of this oscillator is approximately twice that of the oscillator of Fig.6.25. In addition, it offers a lower phase noise

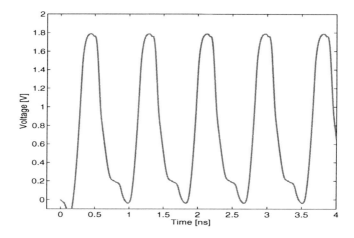

Figure 6.23.   Simulated output voltage of a 8-stage dual-loop Park-Kim VCO.

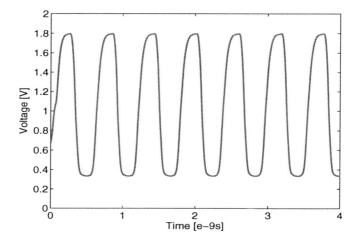

Figure 6.24.   Simulated output voltage of a 8-stage dual-loop Park-Kim VCO with Hara active inductor loads.

as compared with the oscillator of Fig.6.25 [142]. Note that the total current drawn by the oscillator in Fig.6.25 is constant, minimizing the switching noise.

Frequency tuning of monolithic spiral LC oscillators is typically carried out by varying the capacitance of their LC tanks as the inductance tuning of spiral inductors in a monolithic integration is rather difficult. These variable capacitors are normally realized using MOS varactors. As a result, monolithic spiral LC oscillators have a small frequency tuning range, typically around

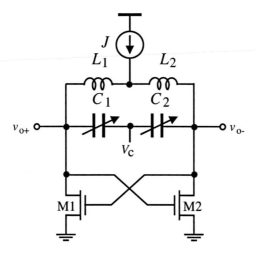

Figure 6.25. LC oscillators with a nMOS negative resistor.

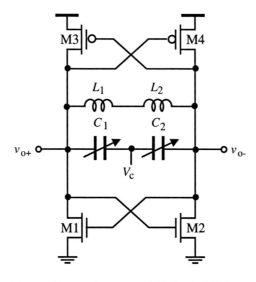

Figure 6.26. LC oscillators with complementary nMOS and pMOS negative resistors.

10%. The requirements for the frequency tuning range in narrow-band wireless applications are much relaxed as compared with those for clock and data recovery in broadband data communications over wire lines and optical channels, varactor-based tuning provides an adequate frequency tuning range. Spiral LC oscillators suffer from a number of drawbacks, in particular,

- Large silicon area - A notable disadvantage of LC oscillators with spiral inductors or transformers is the large silicon area required for the fabrication of the spiral inductors and transformers. Not only the inductance of spiral inductors is low, the silicon area occupied by spiral inductors is also proportional to the inductance of the spiral inductors. Often, the silicon area of a RF front-end is primarily consumed by a few spiral inductors.

- Low self-resonant frequency - The self-resonant frequency of a spiral inductor is set by the inductance of the spiral and the parasitic capacitance between the spiral and the substrate. The large metal area of the spiral gives rise to a large spiral-substrate capacitance, which limits the self-resonant frequency of spiral inductors to a few GHz typically.

- Low quality factor - The ohmic losses of spirals at high frequencies, mainly due to the skin-effect induced resistance of the spirals and the substrate eddy-current induced resistance, limits the quality factor of spiral inductors to below 20 typically. Although the ohmic losses can be compensated for by employing negative resistors, the compensation result is generally poor over a large frequency range. This is because negative resistors are active networks while the parasitic resistance of spirals is a strong function of frequency.

Although active inductors offer several attractive advantages over their spiral counterparts including a large and tunable inductance, a low silicon consumption, and fully realizable in digital-oriented CMOS technologies. The applications of active inductor LC oscillators are confronted with two difficulties : a high level of phase noise and a limited dynamic range. An in-depth research in these areas and novel circuit topologies are critically needed in order to lower the phase noise and improve the dynamic range of these oscillators prior to a large scale deployment of active inductor oscillators in applications where passive inductors are dominating currently.

6.3.1　LC VCOs with Wu Current-Reuse Active Inductors

Fig.6.27 is the simplified schematic of a voltage-controlled LC oscillator employing Wu current-reuse active inductors investigated in Chapter 2. Transistors $M_{1,2,3a/b}$ form two Wu current reuse active inductors whose inductance is tuned by varying the dc biasing current J_1. Transistors $M_{4a/b}$ and J_2 form a negative resistor whose resistance is given by $-\dfrac{2}{g_{m4}}$ approximately and is tunable by varying J_2. This negative resistor is used to cancel out the parasitic resistances of the active inductors.

As demonstrated in [42], the silicon area of the VCO implemented in a $0.35\mu m$ 3V CMOS technology was only $100\mu m \times 120\mu m$. Frequency tuning range of the VCO was from 100 MHz to 900 MHz with phase noise of approximately -95 dBc/Hz at 500 kHz frequency offset. The high level of the phase noise is mainly due to the use of the active inductors and the tail current

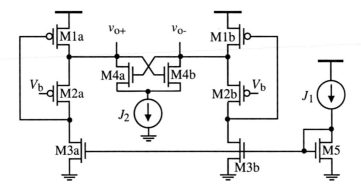

Figure 6.27. Simplified schematic of LC VCO with Wu current-reuse active inductors.

source of the negative resistor. Note that the phase noise can be improved by removing the tail current source of the negative resistor with the downside that the resistance of the negative resistor will not be variable.

6.3.2   LC VCOs with Lin-Payne Active Inductors

Lin-Payne active inductors investigated in Chapter 2 were used in design of LC oscillators. Fig.6.28 shows the simplified schematic of a LC oscillator with Lin-Payne active inductors. The pMOS transistor branches are used as inductance tuning branches subsequently frequency tuning branches while the nMOS transistor branches are connected to a negative resistor to cancel out the parasitic resistances of the active inductor. The resistance of the negative resistor is tuned by varying J_2 for a total compensation. This in turn adjusts the quality factor of the LC oscillator. The inductance of the active inductors is tuned by varying J_1. The quality factor and oscillation frequency of the oscillator can thus be tuned independently.

It was demonstrated in [39] that in a 0.35μm implementation of a LC oscillator with Lin-Payne active inductors, the phase noise of the oscillator when $Q = 7$ was -83 dBc/Hz at 600 kHz frequency offset from a 1.5 GHz oscillation frequency. The power consumption of the oscillator was 2.5 mW. When $Q = 24$, the phase noise was reduced to -88 dBc/Hz at 600 kHz frequency offset from a 2.5 GHz oscillation frequency. The power consumption in this case was 4.5 mW. The frequency tuning range of the oscillator was 48%. The total active area of the oscillator excluding the biasing current sources was only 24 μm^2.

6.3.3   LC VCOs with Grözing Active Inductors

Grözing active inductors investigated in Chapter 2 are differentially configured such that they are less sensitive to common-mode disturbances. They

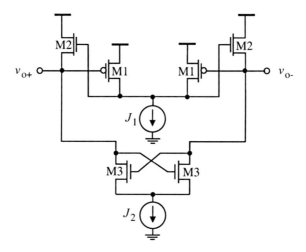

Figure 6.28.   Simplified schematic of LC VCO with Lin-Payne active inductors.

were used in design of LC oscillators in [43, 44]. Fig.6.29 shows the simplified schematic of a LC oscillator with Grözing active inductors. The loads of the two differentially configured transconductors are controlled by V_{b1}. V_{b2} controls the tail biasing currents of the transconductors, which in turn tunes the transconductances of the transconductors, subsequently the inductance of the active inductor and the oscillation frequency of the oscillator. Two negative resistors implemented using the popular cross-coupled nMOS pair are employed at the output nodes of the transconductors to cancel out the parasitic output resistances of the transconductors subsequently the parasitic resistances of the active inductors. Note that the input impedances of the transconductors are infinity. So $G_{o11}, G_{o12}, G_{o21}$ and G_{o22}, which contribute to the parasitic resistances of the active inductor, are solely due to the output resistances of the transconductors. The resistances of the two negative resistors are tuned individually by varying $J_{2,3}$ such that the output impedances of the two transconductors can be compensated for individually.

It was demonstrated in a 0.30 μm 2.5 V implementation of a LC oscillator with Grözing active inductors, the self-resonant frequency of the active inductor was 5.6 GHz [43, 44]. The total power consumption of the oscillator was 15 mW. The quality factor was greater than 100 over the frequency range from 400 MHz to 4 GHz. The layout area of the oscillator was $200\mu m \times 200\mu m$.

6.3.4   LC VCOs with Karsilayan-Schaunann Active Inductors

Karsilayan-Schaunann active inductors investigated in Chapter 2 offer a large quality factor and the independent tunability of the inductance and quality

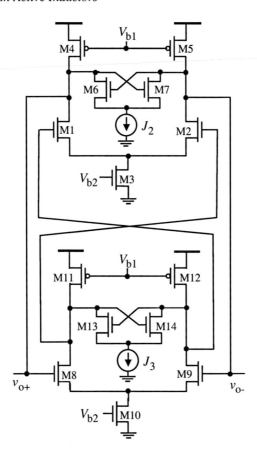

Figure 6.29. Simplified schematic of LC VCO with Grözing active inductors.

factor. LC oscillators that are based on Karsilayan-Schaunann active inductors were developed by Xiao and Schaunann in [75]. The simplified schematic of a LC oscillator employing Karsilayan-Schaunann active inductors is shown in Fig.6.30. Current sources J_{1-6} that provide dc biasing currents can be generated from a single master current source using current mirrors. J_7 is used to tune the resistance of the compensating negative resistor. As pointed out in Chapter 2 that the inductance of Karsilayan-Schaunann active inductors is tuned by varying C_I while its quality factor is adjusted by varying C_Q. These auxiliary capacitors can be implemented using MOS varactors.

It was demonstrated in a 0.20 μm implementation of a LC oscillator with Karsilayan-Schaunann active inductors, the oscillation frequency of the oscillator was 4.95 GHz. The total power consumption was approximately 1.16 mW. The phase noise of the oscillator was -81 dBc/Hz at 500 kHz frequency offset. The total silicon consumption of the oscillator core was $24 \times 22 \mu m^2$.

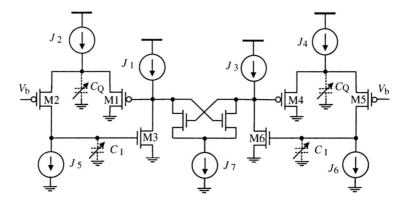

Figure 6.30.   Simplified schematic of LC VCO with Karsilayan-Schaunann active inductors.

6.3.5   LC VCOs with Lu Active Inductors

As compared with differential-pair transconductor-based floating active inductors, Lu differential active inductors investigated in Chapter 2 offer the advantages of a compact configuration subsequently a large self-resonant frequency, a low level of noise, and a low silicon consumption. This inductor was used to construct LC oscillators in [60]. The simplified schematic of a LC oscillator with Lu active inductors is shown in Fig.6.31. M_{1-6} form a Lu floating active inductor. The additional capacitors of the LC tank are constructed using two MOS varactors $M_{7,8}$ whose capacitances are controlled by V_{c2}. They are used for the fine tuning of the oscillation frequency of the oscillator. A negative resistor formed by $M_{9,10}$ is employed to cancel out the parasitic resistances of the active inductor. The inductance of Lu active inductor is tuned by varying V_{c1} with $M_{5,6}$ biased in the triode. V_{c1} is used for the coarse tuning of the oscillation frequency of the oscillator. Note that the resistance of the negative resistor can not be tuned in this configuration. To tune the resistance of the negative resistor, a tail biasing current source that supplies a biasing current to $M_{9,10}$ or a shunt MOS transistor across the drain of $M_{9,10}$, similar to Mahmoudi-Salama floating active inductors investigated in Chapter 2, can be employed.

Implemented in a 0.18μm 1.8V CMOS technology, the LC oscillator with Lu active inductors had a coarse frequency tuning range from 500 MHz to 3.0 GHz [60]. The coarse frequency tuning was done by varying V_{c1} with a sensitivity of 2.5 GHz/V. Fine frequency tuning was achieved by varying V_{c2} with a sensitivity of 108 MHz/V. The measured phase noise of the oscillator was from -101 dBc/Hz to -118 dBc/Hz at 1 MHz frequency offset with the oscillation frequency tunable from 500 MHz to 3.0 GHz. The power consumption of the oscillator was from 6 mW to 28 mW. The silicon area of the oscillator was $150\times300\mu m^2$.

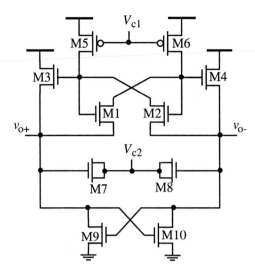

Figure 6.31. Simplified schematic of LC VCO with Lu floating active inductors.

6.4 LC VCOs With Active Transformers

Transformer LC oscillators are the newest members of the VCO family. As compared with spiral inductor LC VCOs, transformer LC VCOs use both the self and mutual inductances of a spiral transformer to form resonating LC tanks, as shown in Fig.6.32 [1]. Transformer LC oscillators offer the critical advantages of a reduced silicon area as compared with LC oscillators with two individual spiral inductors, an increased effective inductance $L_{eff} = L_{11} + kM_{12}$ where k is the coupling coefficient, L_{11} is the self-inductance of the primary winding and M_{12} is the mutual inductance from the secondary winding to the primary winding, and an improved quality factor.

The current of the primary winding I_1 of a lossless transformer generates a voltage drop at the secondary winding of the transformer $V_{21} = (R_{21} + j\omega M_{21})I_1$. Because the transformer does not transfer DC currents from the primary winding to the secondary winding, $R_{21} = 0$. As a result, $V_{21} = j\omega M_{21}I_1$. This observation suggests that the mutual quality factor of the transformer computed from $Q_{21} = \dfrac{\omega L_{21}}{R_{21}}$ is infinite ideally.

Fig.6.33 plots the real and imaginary parts of the input impedance of the primary winding of Wu active inductors investigated in Chapter 2. Because the real part of the impedance is larger than the imaginary part in the frequency range of our interest (>1 GHz), a low self-quality factor of the primary winding exists.

Fig.6.34 shows the real and imaginary parts of the mutual inductance from the primary winding to the secondary winding of Tang active transformers

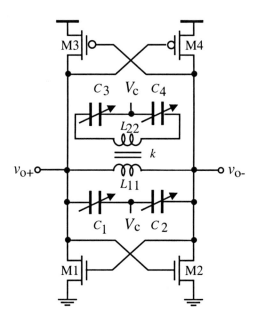

Figure 6.32. LC oscillators with spiral transformers.

investigated in Chapter 3. It is seen that the real part of the mutual inductance vanishes at 1.6 GHz approximately, resulting in a large mutual quality factor at 1.6 GHz.

The large mutual inductance of the active transformer at 1.6 GHz suggests that a lower phase noise can be obtained if LC oscillators are constructed using CMOS active transformers and operate at 1.6 GHz. Fig.6.35 shows the simplified schematic of a LC VCO using Tang active transformers investigated in Chapter 3. The LC VCO is designed to oscillate at 1.6 GHz. Fig.6.36 plots the phase noise of the active transformer oscillator at nominal process condition and at process corners, namely FF (Fast-nMOS/Fast pMOS), FS (Fast nMOS/slow pMOS), SS (slow nMOS/slow pMOS). The phase noise of the VCO is better than that of active inductor LC VCOs investigated earlier in this chapter.

The preceding LC oscillator with Tang active transformers can also be implemented using the class AB active transformers proposed by Tang *et al.* in [64] and investigated in Chapter 3. The schematic of the LC oscillator remains unchanged except that the transformer is replaced with a Tang class AB active transformer. To compare the performance of the preceding LC oscillator with Tang active transformers and Tang class AB active transformers, both oscillators are designed to oscillate at 1.6 GHz and have the same transistor sizes in order to have a fair comparison. The phase noise of the two oscillators is plotted in Fig.6.37. It is seen that the oscillator with Tang class AB transformers

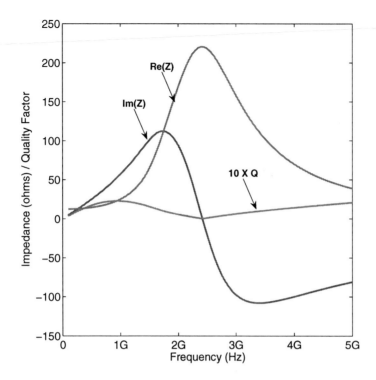

Figure 6.33. Simulated real and imaginary parts of the input impedance of Wu current reuse active inductor. *Q* is multiplied by 10 for a better view (© IEEE 2007).

exhibits a lower level of phase noise. As pointed out in Chapter 3, class AB active transformers exhibit a large average mutual quality factor as compared with their class A counterparts. It is this large mutual quality factor that yields a better phase noise.

Fig.6.38 plots the phase noise of the class AB transformer LC oscillator at four process corners provided by TSMC, namely FF (Fast-nMOS/Fast-pMOS), FS (Fast-nMOS/Slow-pMOS), SF (Slow-nMOS/Fast-pMOS), and SS (Slow-nMOS/Slow-pMOS). It is observed that the phase noise of the oscillator is sensitive to process variation. It should be noted that the biasing condition of the transformer was adjusted for each process corner and supply voltage variation to ensure that the oscillation frequency of the oscillator remains at 1.6 GHz.

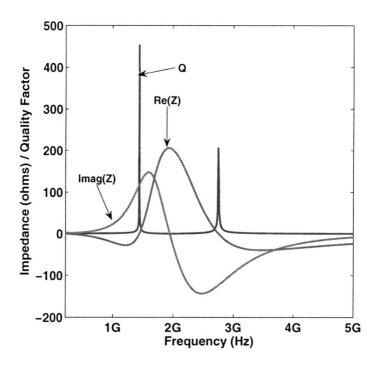

Figure 6.34. Simulated real and imaginary parts of the mutual impedance from the primary winding to the secondary winding of Tang active transformer (© IEEE 2007).

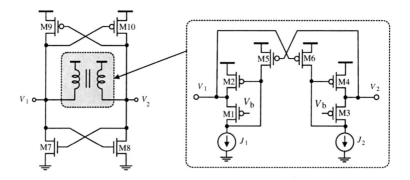

Figure 6.35. LC oscillator with Tang active transformers.

6.5   Quadrature LC VCOs With Active Inductors

Quadrature oscillators are usually implemented using LC-tank VCOs with spiral inductors or transformers. Fig.6.39 shows the simplified schematic of the quadrature LC oscillator proposed in [5]. The simplified schematic of trans-

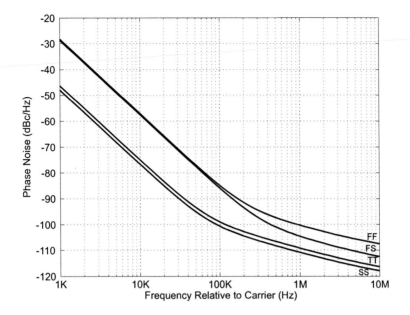

Figure 6.36. Simulated phase noise of LC oscillator with Tang active transformer (© IEEE 2007).

former quadrature oscillator proposed in [2] are shown in Fig.6.40. Variable capacitors C_{1-4} are MOS varactors for frequency tuning.

These oscillators typically require a large silicon area due to the need for four spiral inductors or two spiral transformers. In this section, we show that quadrature oscillators can be implemented using CMOS active inductors to significantly reduce silicon consumption.

Mahmoudi-Salama floating active inductors investigated in Chapter 2 are based on differential-pair transconductors. As compared with Lu floating active inductors, although Mahmoudi-Salama active inductors require more transistors subsequently a larger silicon area, their modular configuration greatly eases the difficulties encountered in both transistor sizing and proper dc biasing. Fig.6.41 shows the simplified schematic of a quadrature LC oscillator with Mahmoudi-Salama active inductors. Transistors $M_{5-7,16-19}$ form negative resistors whose resistances are tuned by varying V_{c2}, as detailed in Chapter 2. MOS varactors are implemented by M_{8-11}. They contribute to the capacitances of the tanks of the oscillator. The oscillation frequency can be tuned by varying V_{c1} for coarse tuning and $V_{c3,c4}$ for fine tuning. Note that $V_{c3,c4}$ also adjust the quadrature phases of the oscillator so that a better quadrature phase relation can be obtained.

Implemented in a 0.18 μm 1.8 V CMOS technology, the quadrature phase and magnitude matching of the oscillator exceeded 1.5 degree and 1 dB, re-

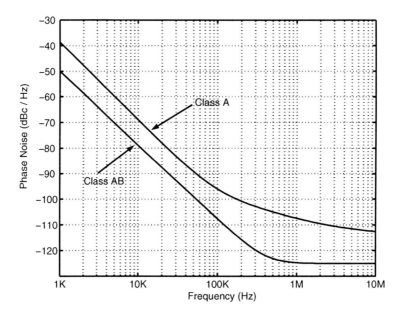

Figure 6.37.   Simulated phase noise of class A and class AB CMOS active transformer VCOs (© IEEE 2007).

spectively over the bandwidth of 100 MHz at 8 GHz. The power consumption of the oscillator was 12 mW. The silicon consumption of the oscillator was $0.15 \times 0.09 \text{mm}^2$.

6.6   Quadrature LC VCOs with Active Transformers

As compared with quadrature LC oscillators with spiral inductors, quadrature LC oscillators with spiral transformers offer several advantages including a smaller silicon area, a large inductance, and an improved quality factor.   In this section, we show that CMOS active transformers investigated in Chapter 3 can be used to construct quadrature LC oscillators to significantly reduce the silicon consumption while providing a good phase noise performance.

The simplified schematic of the quadrature oscillator employing two Tang active transformers is shown in Fig.6.42. The oscillator consists of two LC oscillators with Tang active transformers. Each LC oscillator employs two negative resistors, one implemented using a nMOS pair and the other implemented in a pMOS pair, to improve the quality factor of the transformers, subsequently the phase noise of the oscillator. The coupling between the two tank oscillators is via four nMOS-based transconductors in a parallel manner. The oscillation frequency can be tuned by varying the inductances of the transformers, which is adjustable by varying the dc biasing currents of the transformers.

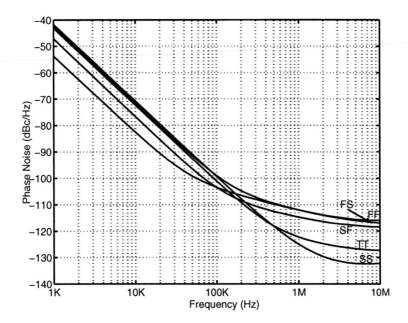

Figure 6.38. Simulated phase noise of class AB transformer oscillator at process corners (© IEEE 2007).

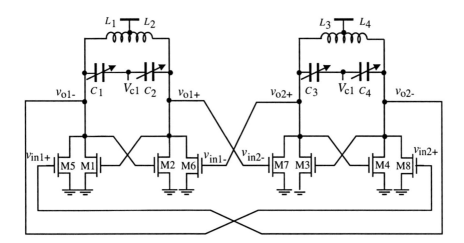

Figure 6.39. Quadrature oscillators with spiral inductors. Two LC oscillators are coupled using parallel transistors M_{5-8}.

Implemented in TSMC-0.18μm 1.8V CMOS technology with an oscillation frequency of 1.6 GHz, the phase noise of the oscillator is plotted in Fig.6.43 in the typical process condition and at process corners. The phase noise in the

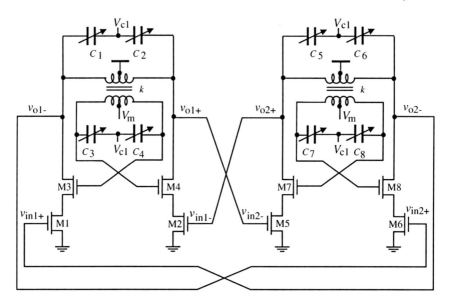

Figure 6.40. Quadrature oscillators with spiral transformers. V_m provides dc biasing voltages for $M_{3,4,7,8}$. Two spiral transformer LC oscillators are coupled using series transistors $M_{1,2,5,6}$.

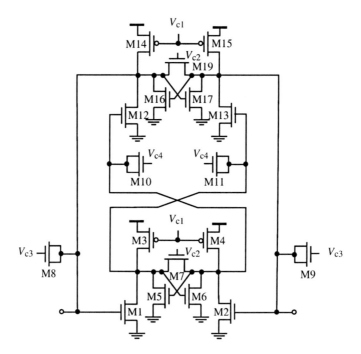

Figure 6.41. LC VCO with Mahmoudi-Salama floating active inductors.

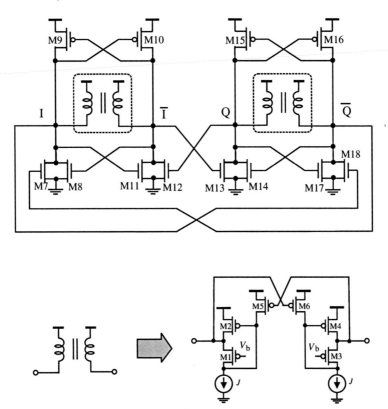

Figure 6.42. Quadrature LC oscillator using Tang active transformers. Two active transformer LC oscillators are coupled using parallel transistors $M_{7,12,13,18}$.

typical process condition is approximately -110 dBc/Hz at 1 MHz frequency offset.

6.7   Performance Comparison of Active LC VCOs

The performance of the active LC oscillators, which include LC oscillators with active inductors and LC oscillators with active transformers is compared in Table 6.1. It is observed that the active LC oscillator with Lu active inductors outperforms other active LC oscillators due to the compact differential configuration of Lu active inductors. The Tang active LC oscillator with active transformers offers a low level of phase noise. The phase noise of the active LC oscillator with Tang class AB active transformers is the lowest (-125 dBc/Hz @ 1 MHz) and is comparable to that of passive LC oscillators. Also the phase noise of the quadrature active LC oscillator proposed by Tang *et al.* offers a low level of phase noise.

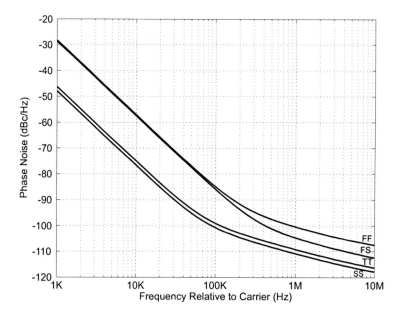

Figure 6.43. Simulated phase noise of quadrature oscillator with Tang active transformers (©) IEEE 2007).

Table 6.1. Performance comparison of active inductor / active transformer LC oscillators.

Ref.	Year	Technology	VCO	f_o [GHz]	Phase noise [dBc/Hz]
Lin-Payne [39]	2000	0.35μm	VCO	2.5	-88 @0.6 MHz
Wu *et al.* [42]	2001	0.35μm	VCO	0.1-0.9	-95 @0.5 MHz
Xiao-Schaumann [75]	2002	0.20μm	VCO	4.95	-81 @0.5 MHz
Lu *et al.*[60]	2006	0.18μm	VCO	0.5-3.0	-101~-118 @1 MHz
Tang *et al.* [63]	2007	0.18μm	TVCO (Class A)	1.6	-108 @1 MHz
Tang *et al.* [64]	2007	0.18μm	TVCO (Class AB)	1.6	-125 @1 MHz
Tang *et al.* [149]	2007	0.18μm	QVCO	1.6	-115 @1 MHz

6.8   Chapter summary

The principles of the operation of ring oscillators and LC oscillators have been presented. We have shown that for LC oscillators, Barkhausen amplitude criterion is satisfied by employing LC tanks as the load of the amplification stages while Barkhausen phase criterion is met by employing two amplification stages. The large impedance of LC tank loads at the resonant frequency of the tanks also ensures that the oscillation frequency of LC oscillators is the resonant frequency of the LC tanks. For ring oscillators, Barkhausen criteria are satisfied by a proper choice of the number of delay stages and the large voltage gain of each delay stage in its transition region. The phase noise of oscillators has also been investigated in detail. We have shown that in order to minimize the phase noise of oscillators, boosting the quality factor of the oscillators is critical.

Three families of ring oscillators with active inductors, namely source-coupled ring oscillators, cross-coupled ring oscillators, and Park-Kim ring oscillators, have been investigated in detail. Source-coupled ring oscillators offer the advantage of a constant current drawn from the supply voltage. As a result, the switching noise generated by the oscillators is minimized. The voltage swing of source-coupled ring oscillators, however, is small due to the voltage drop of the biasing tail current source. Cross-coupled ring oscillators offer a rail-to-rail output voltage swing, however, at the cost of a high level of switching noise. This is because the total current drawn by each delay stage of the oscillators is no longer constant. Both source-coupled and cross-coupled ring oscillators suffer from the drawback of waveform asymmetry because only the charge time of the output nodes of the delay stages of these VCOs is controlled. The key advantage of Park-Kim ring oscillators is the improved waveform symmetry as the delay of the delay stages of these oscillators is controlled by adjusting the intensity of the latch that affects both the charge and discharge times of the stages.

The design of active LC oscillators, which include active inductor LC oscillators, active inductor quadrature LC oscillators, active transformer LC oscillators, and active transformer quadrature LC oscillators, has been investigated in detail. The performance of these active LC oscillators has been compared. We have shown that the phase noise of active LC oscillators is high, approximately 10-20 dB higher as compared with that of passive LC oscillators. The active LC oscillator with Lu active inductors outperforms other active LC oscillators due to the compact differential configuration of Lu active inductors. The Tang active transformer LC oscillator offers a low level of phase noise. The phase noise of the active LC oscillator with Tang class AB active transformers is the lowest (-125 dBc/Hz @ 1 MHz). Also the quadrature active transformer LC oscillator proposed by Tang *et al.* offers a low level of phase noise.

Chapter 7

CURRENT-MODE PHASE-LOCKED LOOPS WITH ACTIVE INDUCTORS & TRANSFORMERS

The main function of a PLL is to align up the phase of its input with that of the output of its local oscillator. To speed up the phase-locking process, a frequency detector that senses the frequency difference between the input of the PLL and the output of its local oscillator is often needed. An introduction of PLLs was available in [68] and a detail treatment of the circuit implementation of the building blocks of PLLs was given in [14]. This chapter focuses on current-mode PLLs with active inductors and transformers. The use of active inductors and active transformers in PLLs offers the critical advantages of a significantly reduced silicon area, a large frequency tuning range of the local oscillator subsequently a large acquisition range, tunable loop dynamics, and a fast locking process.

The chapter is organized as the followings : Section 7.1 reviews the fundamentals of PLLs including classifications, loop dynamics, and phase noise. Section 7.2 deals with active inductor current-mode PLLs. Current-mode loop filters with active inductors are introduced. The loop dynamics of types I and II current-mode PLLs are examined and the design trade-offs are investigated. The phase noise of types I and II current-mode PLLs is also investigated in detail. Two current-mode PLLs, one with an active inductor loop filter and an active inductor LC oscillator and the other with an active inductor loop filter but an spiral inductor LC oscillator, are designed and their performance, in particular, phase noise, is compared. Section 7.3 deals with active transformer current-mode PLLs. Current-mode loop filters with active transformers are introduced and the loop dynamics of current-mode PLLs with active transformers are derived. The phase noise of active transformer current-mode PLLs is investigated. The chapter is summarized in Section 7.4.

7.1 Fundamentals of PLLs

7.1.1 Classifications

Phase-locked loops are often classified on the basis of the order of their closed-loop transfer function. This classification, however, is less convenient from a circuit design perspective as IC designers are usually more familiar with the use of open-loop quantities, such as loop gain, to depict the behavior of closed-loop systems. For example, in analysis of an operational amplifier, a control system designer would be more likely in favor of using a zero / pole approach to determine the stability of the amplifier by examining the location of the zeros and poles of the amplifier. An IC designer, however, would most likely to use the Bodé plots of the loop gain of the amplifier to determine the stability and quantify the phase margin of the amplifier. The zero/pole approach, though exact and effective, is often difficult for high-order systems. The Bodé plot approach, on the other hand, is particularly efficient in determination of the stability of high-order systems.

Depending upon the number of the poles of the loop gain of PLLs at the origin ($s = 0$), PLLs can be classified into type I and type II. A PLL is said to be type I if its loop gain has only one pole at the origin. It is a type II PLL if its loop gain has two poles at the origin. High-order PLLs are rarely used in practice due to stability concerns. The different number of the poles of the loop gain of PLLs at the origin gives rise to the distinct behavior of the PLLs.

Phase-locked loops can also be classified on the basis of the type of the control signal of their local oscillator. A PLL is said to be voltage-mode if the oscillation frequency of its local oscillator is voltage-controlled. A typical configuration of voltage-mode PLLs is shown in Fig.7.1. A PLL is termed current-mode if the oscillation frequency of its local oscillator is current-controlled. Fig.7.2 gives the configuration of current-mode PLLs. Note that a buffer is usually needed at the output of the oscillator to ensure that the feedback signal is Boolean as phase/frequency detectors are typically digital blocks. The buffer also minimizes the effect of the load of the oscillator on the performance of the oscillator

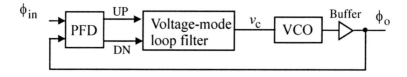

Figure 7.1. Configuration of voltage-mode PLLs.

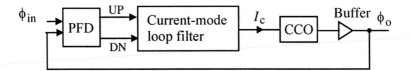

Figure 7.2. Configuration of current-mode PLLs.

7.1.2 Loop Dynamics of Voltage-Mode PLLs

When a PLL with a linear phase detector, such as DFF-based phase-frequency detector, is in the lock state, the variation of the control voltage or current of its local oscillator is small. In this case, the PLL can be treated as a linear system approximately and its behavior can be analyzed using the widely used s-domain approach for linear feedback control systems. For PLLs with bang-bang phase detectors, the analysis error of using the s-domain approach will arise because of the large fluctuation of the control voltage or current of the oscillator.

Loop Dynamics of Type I Voltage-Mode PLLs

The output of a linear phase detector of a voltage-mode PLL is a train of pulses whose width is the same as the phase difference between its two inputs. The average voltage of the phase detector over an oscillation period is directly proportional to the phase difference. The behavior of linear phase detectors in the s-domain is depicted by [68]

$$H_{pd}(s) = K_{pd}, \tag{7.1}$$

where K_{pd} is the gain of the phase detector. The relation between the variation of the excess phase of a voltage-controlled oscillator and its control voltage is depicted using an integrator [68]

$$H_{vco}(s) = \frac{K_{vco}}{s}, \tag{7.2}$$

where K_{vco} is the gain of the VCO. If the low-pass loop filter of the PLL has the transfer function

$$H_{LF}(s) = \frac{1}{\dfrac{s}{\omega_{LF}} + 1}, \tag{7.3}$$

where ω_{LF} is the bandwidth of the loop filter, the open-loop gain of the PLL is given by

$$H_o(s) = H_{pd}(s)H_{LF}(s)H_{vco}(s) = \frac{K_{pd}}{1 + \dfrac{s}{\omega_{LF}}} \frac{K_{vco}}{s}. \qquad (7.4)$$

It is evident from (7.4) that the PLL is type I as $H_o(s)$ has only one pole at $s = 0$. Notice that $H_o(s)\big|_{s=0} = \infty$ and the closed-loop gain of the PLL becomes

$$H_c(s)\big|_{s=0} = \frac{H_o(s)}{1 + H_o(s)}\bigg|_{s=0} = 1. \qquad (7.5)$$

Eq.(7.5) reveals that low-frequency noise encountered at the input of a type I PLL will appear at the output of the PLL without any attenuation. The closed-loop transfer function of the PLL is given by

$$H_c(s) = \frac{K_{pd}K_{vco}\omega_{LF}}{s^2 + s\omega_{LF} + K_{pd}K_{vco}\omega_{LF}}. \qquad (7.6)$$

The pole resonant frequency, also known as the loop bandwidth ω_n, and the damping factor ξ of the PLL can be obtained from (7.6) directly [150]

$$\omega_n = \sqrt{K_{pd}K_{vco}\omega_{LF}},$$

$$\xi = \frac{1}{2}\sqrt{\frac{\omega_{LF}}{K_{pd}K_{vco}}}. \qquad (7.7)$$

From (7.7) we have

$$\omega_n\xi = \frac{\omega_{LF}}{2}. \qquad (7.8)$$

The poles of $H_c(s)$ are obtained from solving its characteristic equation

$$\begin{aligned} s_{1,2} &= \frac{\omega_{LF}}{2}\left(-1 \pm \sqrt{1 - \frac{1}{\xi^2}}\right) \\ &= \omega_n\xi\left(-1 \pm \sqrt{1 - \frac{1}{\xi^2}}\right). \end{aligned} \qquad (7.9)$$

We comment on the preceding results :

- The poles of the PLL, $s_{1,2}$, are a pair of complex conjugates with a negative real part, revealing that the PLL is stable. The response of the PLL is an oscillating quantity with oscillation frequency $\omega_n\xi\sqrt{1 - \dfrac{1}{\xi^2}}$ and decaying amplitude $e^{-\omega_n\xi t} = e^{-\frac{\omega_{LF}}{2}t}$.

- An increase in ω_{LF} will result in an increase in the real part of $s_{1,2}$. As a result, the poles will move further away from the imaginary axis and the PLL will be more stable. In addition, the oscillation frequency of the output of the PLL will increase and the amplitude of the oscillating output of the PLL will decrease. This is echoed with a reduced lock time of the PLL.

- An increase in ξ will increase the real part and decrease the imaginary part of $s_{1,2}$. As a result, the attenuation of the output of the PLL will increase and the frequency of the oscillating output of the PLL will decrease.

- ω_n and ξ can not be tuned independently. Any change of K_{pd}, K_{vco}, or ω_{LF} will affect both ω_n and ξ. This is a drawback of type I voltage-mode PLLs.

Loop Dynamics of Type II Voltage-Mode PLLs

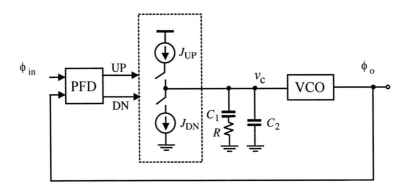

Figure 7.3. Configuration of type II voltage-mode charge-pump PLLs.

The configuration of charge-pump PLLs is shown in Fig.7.3. Consider the case where the loop filter is implemented using C_1 only (i.e. neglect R and C_2). Because the transfer function of the loop filter in this case becomes

$$H_{LF}(s) = \frac{1}{sC_1},\qquad (7.10)$$

the open-loop gain has two poles at the origin and the PLL is type II. Further because the closed-loop transfer function of the PLL has two poles on the imaginary axis, the PLL is unstable.

To stabilize of the system, a zero is needed. This can be realized by adding a resistor R in series with C_1, as shown in Fig.7.3 (without C_2). The transfer function of the loop filter is given by

$$H_{LF}(s) = \frac{V_c(s)}{I_{cp}(s)} = \frac{sRC_1 + 1}{sC_1}. \tag{7.11}$$

The loop gain of the PLL becomes

$$H_o(s) = K_{pd}\left(\frac{sRC_1 + 1}{sC_1}\right)\left(\frac{K_{vco}}{s}\right). \tag{7.12}$$

The PLL is still type II. From the closed-loop transfer function of the PLL

$$H_c(s) = \frac{\Phi_o(s)}{\Phi_{in}(s)} = \frac{K_{pd}K_{vco}R\left(s + \dfrac{1}{RC_1}\right)}{s^2 + sK_{pd}K_{vco}R + \dfrac{K_{pd}K_{vco}}{C_1}}, \tag{7.13}$$

we obtain the loop bandwidth ω_n and damping factor ξ of the PLL

$$\omega_n = \sqrt{\frac{K_{pd}K_{vco}}{C_1}}, $$
$$\xi = \frac{R}{2}\sqrt{K_{pd}K_{vco}C_1}. \tag{7.14}$$

Also,

$$\omega_n\xi = \frac{1}{2}RK_{pd}K_{vco}. \tag{7.15}$$

The two poles of $H_c(s)$ are located at

$$\begin{aligned} s_{1,2} &= \frac{1}{2}K_{pd}K_{vco}R\left(-1\pm\sqrt{1 - \frac{4}{K_{pd}K_{vco}R^2C_1}}\right) \\ &= \omega_n\xi\left(-1\pm\sqrt{1 - \frac{1}{\xi^2}}\right). \end{aligned} \tag{7.16}$$

It is seen that both poles are located in the left half of the s-plane and the PLL is stable. Critical damping occurs when the system has two identical real poles at which $K_{pd}K_{vco}R^2C_1 = 4$. Also notice that the location of the poles of $H_c(s)$ is the same as that of the preceding type I PLL, revealing that both PLLs have the same stability property. We further comment on the preceding results :

- If we use the real part of the poles as a measure of the absolute stability of the PLL, an increase of R will result in an increase in the absolute value of the real part of the poles $s_{1,2}$, i.e. the poles will move further away from the imaginary axis and the system becomes more stable.

- The imaginary part of the poles $s_{1,2}$ can be used as a measure of the transient response time of the system. The larger the imaginary part of the poles, the faster the response of the system to a change in the input of the PLL. It is evident from (7.16) that an increase in the loop bandwidth ω_n will increase the imaginary part of $s_{1,2}$, reflected by a decrease in the lock time of the PLL. The same conclusion can be drawn for ω_{LF} as well. An increase in ω_n will also speed up the attenuation of the amplitude of the oscillating output of the PLL.

- An increase of C_1 will reduce the imaginary part of $s_{1,2}$ and lower the loop bandwidth. This is echoed with a reduction in the lock time of the PLL. Note that C_1 does not affect $\Re e[s_{1,2}]$, i.e. the absolute stability of the system remains unchanged.

- Unlike type I voltage-mode PLLs studied earlier, the damping factor ξ of type II voltage-mode PLLs can be tuned by varying R without affecting the loop bandwidth ω_n. This is an important advantage of type II voltage-mode PLLs over their type I counterparts.

- To suppress the noise injected at the input of the PLLs, the loop bandwidth of the PLLs must be made small. For a given damping factor, the bandwidth of the loop filter ω_{LF} must also be made small. As a result, the value of C_1 is large. Because C_1 of a large capacitance behaves as a short-circuit for high-frequency signals on the control line, these high-frequency signals generated mainly by the preceding charge pump due to their nonidealities will create a voltage drop across the resistor. They will appear in the spectrum of the output of the PLLs as reference spurs, deteriorating the phase noise of the PLLs. To minimize this unwanted effect, a small capacitor C_2 can be connected in parallel with the main RC_1 branch, as shown in Fig.7.3, to shunt these high-frequency disturbances to the ground. For low frequency signals, C_2 behaves as an open-circuit.

Table 7.1 compares the characteristics of type I and type II voltage-mode PLLs.

Table 7.1. Comparison of the performance of type I and type II voltage-mode PLLs.

Design parameters	Type I PLLs	Type II PLLs
Loop bandwidth (ω_n)	$\sqrt{K_{pd}K_{vco}\omega_{LF}}$	$\sqrt{\dfrac{K_{pd}K_{vco}}{C_1}}$
Damping factor (ξ)	$\dfrac{1}{2}\sqrt{\dfrac{\omega_{LF}}{K_{pd}K_{vco}}}$	$\dfrac{1}{2}R\sqrt{K_{pd}K_{vco}C_1}$
Loop bandwidth-damping factor product ($\omega_n\xi$)	$\dfrac{1}{2}\omega_{LF}$	$\dfrac{1}{2}RK_{pd}K_{vco}$
Poles ($s_{1,2}$)	$\dfrac{\omega_{LF}}{2}\left(-1\pm\sqrt{1-\dfrac{1}{\xi^2}}\right)$ $=\omega_n\xi\left(-1\pm\sqrt{1-\dfrac{1}{\xi^2}}\right)$	$\dfrac{\omega_{LF}}{2}\left(-1\pm\sqrt{1-\dfrac{1}{\xi^2}}\right)$ $=\omega_n\xi\left(-1\pm\sqrt{1-\dfrac{1}{\xi^2}}\right)$

7.1.3   Phase Noise of Voltage-Mode PLLs

To analyze the phase noise of voltage-mode PLLs in the lock state, consider the type II voltage-mode PLL shown in Fig.7.4 where N_{in}, N_{LF}, and N_{vco} denote the noise from the input, the loop filter, and the voltage-controlled oscillator, respectively. The transfer functions from these noise sources to the output of the PLL are given by

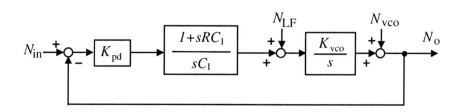

Figure 7.4.   Noise analysis of type II voltage-mode PLLs.

$$H_{vco}(s) = \frac{N_o(s)}{N_{vco}(s)} = \frac{s^2}{s^2 + 2\xi\omega_n\omega + \omega_n^2},$$

$$H_{in}(s) = \frac{N_o(s)}{N_{in}(s)} = \frac{\omega_n^2(\dfrac{s}{\omega_{LF}} + 1)}{s^2 + 2\xi\omega_n\omega + \omega_n^2}, \qquad (7.17)$$

$$H_{LF}(s) = \frac{N_o(s)}{N_{LF}(s)} = \frac{K_{vco}s}{s^2 + 2\xi\omega_n\omega + \omega_n^2}.$$

It is seen from (7.17) that the transfer characteristic from the noise source of the VCO to the output of the PLL is high-pass with the corner frequency ω_n, the transfer characteristic from the loop filter to the output of the PLL is band-pass with its center frequency ω_n, and the transfer characteristic from the input of the PLL to the output of the PLL is low-pass with its cutoff frequency ω_n, as illustrated graphically in Fig.7.5. This observation reveals that the noise from the VCO contributes the most to the phase noise of the PLL at frequencies beyond ω_n, the noise from the input of the PLL contributes the most to the phase noise of the PLL at frequencies below ω_n , and the noise from the loop filter contributes the most to the phase noise of the PLL in the vicinity of ω_n.

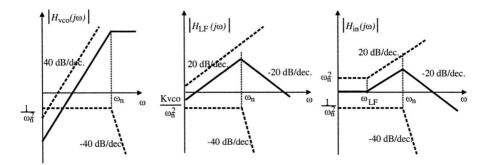

Figure 7.5. Bodé plot (magnitude) of the transfer functions from the noise sources to the output of type II voltage-mode PLLs.

In order to suppress the noise originated from the VCO, the loop bandwidth ω_n should be maximized, as is evident in Fig.7.6. The minimization of the noise injected at the input of the PLL requires that ω_n be minimized, as is evident in Fig.7.7.

The distinct noise transfer characteristics of PLLs require that the loop characteristics of PLLs be tailored for specific applications. As an example, the input of the PLLs of the transmitter of a serial link is usually from a low-noise crystal oscillator. In this case, maximizing the loop bandwidth of the

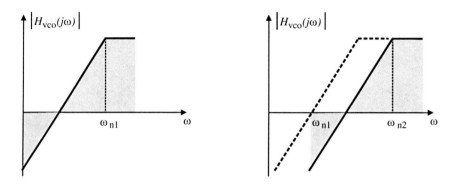

Figure 7.6. Bodé plot (magnitude) of the transfer function from the noise source of the VCO to the output of type II voltage-mode PLLs ($\omega_{n2} > \omega_1$)..

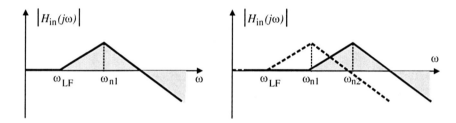

Figure 7.7. Bodé plot (magnitude) of the transfer function from the noise source of the input to the output of type II voltage-mode PLLs ($\omega_{n2} > \omega_1$).

PLL is critical to suppress the noise generated by the VCO. The input of the PLL in the receiver of a serial link is the received data stream that contains a high level of timing jitter, the loop bandwidth of the PLLs in this case should be made small to suppress the noise from the incoming data stream.

7.1.4   Simulation of Phase Noise of PLLs

Unlike oscillators, a stiff difficulty exists in analysis of the phase noise of PLLs. The difficulty arises due to the following reasons :

- Unlike oscillators whose oscillation frequency is fixed at a given control voltage or current, PLLs are not periodically time-varying systems. The analysis of the phase noise of oscillators is based on the principle of cyclostationary stochastic processes where stationary noise, such as thermal noise of MOS devices, is sampled in a periodical fashion. Due to the variation of the control voltage of voltage-mode PLLs, even in the lock state, the oscillation frequency of the oscillator of the PLLs is not constant.

- Phase-locked loops contains harsh nonlinear elements, such as phase detectors. Simulation of PLLs using analog simulators in the time domain, such as SpectreRF, is extremely time-consuming.

Current CAD tools for IC design from leading CAD vendors, such as SpectreRF from Cadence Design Systems, can not analyze the phase noise of arbitrary PLLs. For PLLs with a linear phase detector, because the variation of the control voltage or current in the lock state is typically small, the variation of the oscillation frequency of the oscillator is small. As a result, the PLLs in the lock state can be considered as periodically time-varying systems approximately. In this case, SpectreRF can be used to estimate the phase noise of the PLLs [151].

A more reliable way to analyze the phase noise of PLLs is to use Verilog-AMS, a hardware description language for mixed analog-digital circuits from Cadence Design Systems [152, 153, 151] to perform time-domain analysis of PLLs with the noise from the building blocks of PLLs considered. Specifically,

- A schematic-level simulation of the building blocks of a PLL is carried out using SpectreRF. The power of the input-referred noise generators of these building blocks is obtained. The noise of the building blocks is then represented by a pseudo-random noise generator in the time with its power the same as that obtained from SpectreRF.

- The phase noise of the CCO was analyzed using Cadence's SpectreRF with the consideration of the fold-over of the broad-band noise sources. The amplitude of the timing jitter of the CCO was obtained from its phase noise. In Verilog-AMS time-domain simulation of the PLL, the oscillation period of the CCO was disturbed using the extracted timing jitter, together with a normally distributed random generator with zero mean and unity variance.

- A time-domain analysis of the PLL is then carried out and the transient response of the PLL is obtained. The data of the steady-state portion of the time-domain response of the PLL is then processed using FFT to yield the spectrum of the power of the phase noise of the PLL.

7.2 Current-Mode PLLs with Active Inductors

This section focuses on the design of current-mode PLLs where active inductors are used in both the loop filter and LC oscillator of the PLLs to reduce the silicon consumption and to provide both tunable loop dynamics and a large acquisition range.

7.2.1 Current-Mode Loop Filter with Active Inductors

As pointed out earlier that a PLL is classified as current-mode if the oscillation frequency of the oscillator of the PLL is current-controlled. The control current is generated from a current-mode loop filter. Current-mode filtering differs fundamentally from voltage-mode filtering. The former sustains a constant output current while the latter outputs a constant output voltage. Voltage-mode low-pass filters are typically constructed using RC networks with shunt capacitors. These capacitors behave as short-circuits at high frequencies to shunt high-frequency signals to the ground. A typical current-mode low-pass filter is comprised of floating inductors, as shown in Fig.7.8, to block high-frequency currents from passing through the inductors.

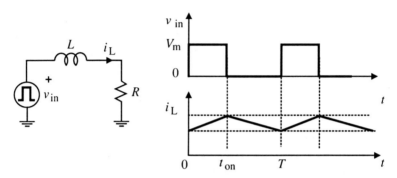

Figure 7.8.   Basic configuration of current-mode low-pass filters.

Let the input voltage of the filter, $v_{in}(t)$, be a square-wave generator with amplitude V_m and a duty cycle t_{on}/T, where T is the period and t_{on} is the ON time of the input. The inductor current is given by

$$
i_L(t) = \begin{cases} \dfrac{V_m}{R}(1 - e^{-t/\tau}) + i_L(0^-)e^{-t/\tau}, & 0 \le t \le t_{on}, \\[3mm] i_L(t_{on}^-)e^{-(t-t_{on})/\tau}, & t_{on} \le t \le T, \end{cases}
\tag{7.18}
$$

where

$$
\tau = \frac{L}{R} \tag{7.19}
$$

is the time constant, $i_L(0^-)$ and $i_L(t_{on}^-)$ are the initial current of the inductor at $t = 0$ and $t = t_{on}$, respectively. If τ is sufficiently large, we have

$$
e^{-\frac{t}{\tau}} \approx 1 - \frac{t}{\tau}. \tag{7.20}
$$

Eq.(7.18) is simplified to

$$i_L(t) \approx \begin{cases} \dfrac{V_m}{R}\dfrac{t}{\tau} + i_L(0^-)\left(1 - \dfrac{t}{\tau}\right), & 0 \leq t \leq t_{on}, \\[2ex] i_L(t_{on}^-)\left(1 - \dfrac{t - t_{on}}{\tau}\right), & t_{on} \leq t \leq T. \end{cases} \qquad (7.21)$$

If the amplitude of the input voltage is kept constant, the dc component of the current-mode filter is a function of the duty cycle of the input voltage of the filter. The output of a linear phase detector is a train of pulses whose pulse width is directly proportional to the phase difference between the inputs of the phase detector. This observation reveals that the dc component of the output current of the current-mode loop filter that follows a linear phase detector is proportional to the phase difference between the inputs of the phase detector.

Low-pass current-mode loop filters require the use of floating inductors to block high-frequency components of their inputs so that the ripple of the control current is small. Although these floating inductors can be implemented using spiral inductors, the need for a large silicon area to fabricate these spiral inductors increases the cost substantially. Floating active inductors investigated in Chapter 2 can be employed as an alternative to construct current-mode loop filters. Three main concerns of this approach, however, exist :

- Active inductors exhibit a significantly high level of noise as compared with their spiral counterparts. The effect of the noise of the active inductors on the overall phase noise of PLLs must be considered.

- The inductance of active inductors are sensitive to supply voltage fluctuation and ground bouncing. On the contrary, the performance of spiral inductors is ideally insensitive to supply voltage fluctuation and ground bouncing. The effect of the fluctuation of the supply voltage and that of ground bouncing on the bandwidth of the active inductor loop filters subsequently the phase noise of PLLs must be considered.

- Active inductor loop filters consume static power.

Because the larger the inductance of the loop filter inductor, the lower the ripple of its output current, floating active inductors of current-mode loop filters of PLLs must therefore have a large inductance. As pointed out earlier in Chapter 2, the inductance of active inductors is inversely proportional to the product of the transconductances of the transconductors constituting the active inductors. A large inductance can therefore be obtained by lowering the dc biasing current of the transconductors. This in turn lowers the power consumption of the active inductors. This observation suggests that the power consumption of active inductor loop filters is not of a critical concern.

7.2.2   Loop Dynamics of Type I Current-Mode PLLs

In what follows we examine the loop dynamics of current-mode PLLs in the lock state. To simplify analysis, we first consider the case where the inductors of current-mode loop filters are assumed to be ideal. We then examine the loop dynamics of the PLLs with the inclusion of the parasitic resistances and capacitance of practical active inductors.

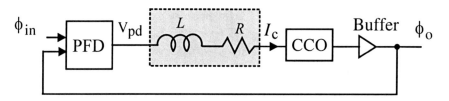

Figure 7.9.   Type I current-mode PLLs with an ideal inductor loop filter.

The configuration of type I current-mode PLLs is shown in Fig.7.9. To simplify analysis, we first neglect the series resistor R. Because the input of the current-controlled oscillator is a current, the input impedance of the current-controlled oscillator can be considered to be zero ideally. As a result, the transfer function of the current-mode loop filter becomes

$$\frac{I_c(s)}{V_{pd}(s)} = \frac{1}{sL}. \tag{7.22}$$

The closed-loop transfer function of the PLL is given by

$$H_c(s) = \frac{\Phi_o(s)}{\Phi_{in}(s)} = \left(\frac{K_{pd}K_{cco}}{L}\right)\frac{1}{s^2 + \dfrac{K_{pd}K_{cco}}{L}}, \tag{7.23}$$

where K_{cco} is the gain of the CCO. The two poles of $H_c(s)$ are on the imaginary axis and the PLL is unstable. To stabilize the system, a resistors R can be added in series with the inductor. The transfer function of the current-mode loop filter in this case becomes

$$\frac{I_c(s)}{V_{pd}(s)} = \frac{1}{sL + R} \tag{7.24}$$

from which we obtain the cut-off frequency of the loop filter

$$\omega_{LF} = \frac{R}{L}. \tag{7.25}$$

The loop gain of the PLL given by

$$H_o(s) = \left(\frac{K_{pd}}{sL + R}\right)\left(\frac{K_{cco}}{s}\right) \tag{7.26}$$

reveals that the PLL is type I as $H_o(s)$ only has one pole at the origin. The closed-loop transfer function of the PLL is given by

$$H_c(s) = \left(\frac{K_{pd}K_{cco}}{L}\right)\frac{1}{s^2 + s\dfrac{R}{L} + \dfrac{K_{pd}K_{cco}}{L}}. \tag{7.27}$$

It is evident from (7.27) that the added resistor moves the poles of the PLL from the imaginary axis to the left half of the s-plane and stabilizes the PLL. The loop bandwidth ω_n and damping factor ξ of the PLL can be derived from (7.27) directly

$$\omega_n = \sqrt{\frac{K_{pd}K_{cco}}{L}},$$

$$\xi = \frac{R}{2}\sqrt{\frac{1}{K_{pd}K_{cco}L}}, \tag{7.28}$$

from which we obtain

$$\omega_n\xi = \frac{\omega_{LF}}{2}. \tag{7.29}$$

An increase in L will lower ω_n and ξ. As a result, the PLL is less stable. The attenuation of the oscillating amplitude of the output of the PLL is also reduced, resulting in a longer lock time. The poles of $H_c(s)$ are given by

$$p_{1,2} = \frac{\omega_{LF}}{2}\left(-1\pm\sqrt{1 - \frac{1}{\xi^2}}\right). \tag{7.30}$$

The location of the poles of the PLL is the same as that of corresponding voltage-mode PLLs derived earlier, revealing that the stability property of current-mode PLLs is the same as that of voltage-mode PLLs.

The damping factor ξ of the PLL can be tuned by varying R without affecting the loop bandwidth. Recall that the loop bandwidth and damping factor of type

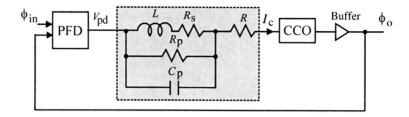

Figure 7.10.   Type I current-mode PLLs with an active inductor loop filter.

I voltage-mode PLLs can not be tuned individually. The independent tunability of ω_n an ξ is a distinct and attractive characteristic of type I current-mode PLLs.

The preceding analysis assumed that the active inductor of the loop filter is ideal. Practical active inductors have both parasitic resistances R_p and R_s, and parasitic capacitance C_p, as shown in Fig.7.10. Their effect on the loop dynamics of current-mode PLLs should be accounted for. Because typically $R_p \gg R$, $R \gg R_s$, R_s can be neglected in analysis of the characteristics of the loop filter without introducing a large error. The closed-loop transfer function of the PLL with the exclusion of C_p is given by

$$H_c(s) = \left[\frac{K_{pd}K_{cco}}{L(R_p + R)}\right] \frac{sL + R_p}{s^2 + s\left(\dfrac{R}{L} + \dfrac{K_{pd}K_{cco}}{R_p}\right) + \dfrac{K_{pd}K_{cco}}{L}\left(\dfrac{R_p}{R + R_p}\right)} \quad (7.31)$$

The loop bandwidth and damping factor in this case are given by

$$\hat{\omega}_n = \omega_n \sqrt{\frac{R_p}{R_p + R}}, \quad (7.32)$$

$$\hat{\xi} = \frac{R}{2}\sqrt{\frac{1}{K_{pd}K_{cco}L}} + \frac{1}{2R_p}\sqrt{K_{pd}K_{cco}L}.$$

Making use of

$$\sqrt{\frac{1}{1+x}} = 1 - \frac{1}{2}x + \frac{3}{8}x^2 - \ldots \approx 1 - \frac{1}{2}x, \quad (x < 1), \quad (7.33)$$

and noting that $\dfrac{R}{R_p} \ll 1$, we have

$$\sqrt{\frac{R_p}{R_p + R}} = \sqrt{\frac{1}{1 + \dfrac{R}{R_p}}} \approx 1 - \frac{1}{2}\frac{R}{R_p}. \quad (7.34)$$

The loop gain becomes

$$\hat{\omega}_n \approx \omega_n \left(1 - \frac{1}{2}\frac{R}{R_p}\right) = \omega_n + \Delta\omega_n. \tag{7.35}$$

Also, the damping factor is given by

$$\hat{\xi} = \xi + \Delta\xi. \tag{7.36}$$

The variations of ω_n and ξ, denoted by $\Delta\omega_n$ and $\Delta\xi$, respectively, which are caused by the parasitics of active inductors, are quantified by

$$\Delta\omega_n = -\frac{1}{2}\frac{R}{R_p}\omega_n,$$

$$\Delta\xi = \frac{1}{4\xi}\frac{R}{R_p}, \tag{7.37}$$

where ω_n and ξ are the loop bandwidth and damping factor of the PLL when the inductor of the loop filter is ideal. It becomes evident from (7.37) that the loop bandwidth is reduced and the damping factor is increased slightly when the parasitic resistances of the active inductor are accounted for. The variations of ω_n and ξ are directly proportional to $\dfrac{R}{R_p}$.

When C_p is considered, because C_p is in parallel with R_p, the phase transfer function in this case is given by (7.31) with R_p replaced with $Z_p = R_p||\dfrac{1}{sC_p}$. Eq.(7.37) can still be used to quantify $\Delta\omega_n$ and $\Delta\xi$ with R_p replaced with Z_p.

7.2.3 Loop Dynamics of Type II Current-Mode PLLs

The configuration of type II current-mode PLLs is shown in Fig.7.11 where the loop is stabilized by adding a resistor in parallel with the inductor.

Following the same assumption that the input impedance of the CCO is zero, we obtain the transfer function of the loop filter

$$\frac{I_c(s)}{V_{pd}(s)} = \frac{sL + R}{sRL}. \tag{7.38}$$

The loop gain of the PLL given by

$$H_o(s) = \frac{K_{pd}K_{cco}(sL + R)}{s^2RL} \tag{7.39}$$

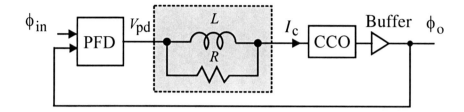

Figure 7.11.   Type II current-mode PLLs.

indicates that the PLL is type II as the loop gain has two poles at the origin. The closed-loop transfer function of the PLL is given by

$$H_c(s) = \frac{K_{pd}K_{cco}}{RL} \frac{sL + R}{s^2 + s\left(\dfrac{K_{pd}K_{cco}}{R}\right) + \dfrac{K_{pd}K_{cco}}{L}}. \tag{7.40}$$

The loop bandwidth ω_n and damping factor ξ of the PLL are obtained from (7.40)

$$\omega_n = \sqrt{\frac{K_{pd}K_{cco}}{L}},$$

$$\xi = \frac{1}{2R}\sqrt{K_{pd}K_{cco}L} \tag{7.41}$$

from which we obtain

$$\omega_n \xi = \frac{1}{2R}K_{pd}K_{cco}. \tag{7.42}$$

The poles of $H_c(s)$ are given by

$$s_{1,2} = \omega_n \xi \left(-1 \pm \sqrt{1 - \frac{1}{\xi^2}}\right). \tag{7.43}$$

We comment the preceding results :

- The locations of the poles of type II current-mode PLL are the same as those of type I current-mode PLLs, revealing that types I and II PLLs have the same stability property.

- An increase in the inductance L will reduce the loop bandwidth ω_n and increase the damping factor ξ. This differs from type I current-mode PLLs where an increase of L will lower both ω_n and ξ.

- An increase in R will lower the damping factor ξ. This is different from type I current-mode PLLs where an increase in R will increase the damping factor ξ.

- The expression of the loop bandwidth ω_n is the same for both types I and II current-mode PLLs, revealing that ω_n is independent of R, both its value and location in the loop filter.

- The resistor R provides a direct path for disturbances to bypass the inductor and reach the CCO, deteriorating the phase noise of the CCO. To eliminate this drawback, an inductor can be added in series with the resistor to block high-frequency currents, as shown in Fig.7.12.

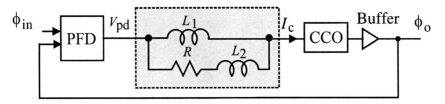

Figure 7.12. Type II current-mode PLLs.

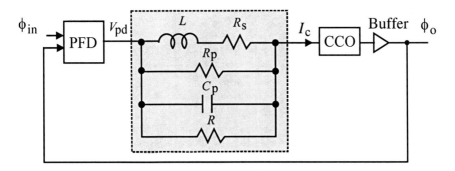

Figure 7.13. Type II current-mode PLLs with an active inductor loop filter.

Let us now repeat the preceding analysis of type II current-mode PLLs with the ideal inductor of the current-mode loop filter replaced with an active inductor that has both parasitic resistances and capacitance, as shown in Fig.7.13. Neglect R_p as $R_p \gg R$. The closed-loop transfer function of the PLL with C_p neglected is given by

$$H_c(s) = \left(\frac{K_{pd}K_{cco}}{LR_p} \right) \frac{sL + (R + R_s)}{s^2 + s\left(\dfrac{R_s}{L} + \dfrac{K_{pd}K_{cco}}{R} \right) + \dfrac{K_{pd}K_{cco}}{L}\left(1 + \dfrac{R_s}{R} \right)}.$$

$$(7.44)$$

The loop bandwidth and damping factor are given by

$$\hat{\omega}_n = \omega_n \sqrt{1 + \frac{R_s}{R}},$$

$$(7.45)$$

$$\hat{\xi} \approx \frac{R_s}{2} \sqrt{\frac{1}{K_{pd}K_{cco}L}} + \frac{1}{2R}\sqrt{K_{pd}K_{cco}L},$$

where ω_n is given in (7.41). Making use of

$$\sqrt{1+x} = 1 + \frac{1}{2}x - \frac{1}{8}x^2 + \ldots \approx 1 + \frac{1}{2}x, \qquad (x < 1), \qquad (7.46)$$

and noting that $R_s \ll R$, we have

$$\sqrt{1 + \frac{R_s}{R}} \approx 1 + \frac{1}{2}\frac{R_s}{R}. \qquad (7.47)$$

As a result

$$\hat{\omega}_n \approx \omega_n\left(1 + \frac{1}{2}\frac{R_s}{R}\right) = \omega_n + \Delta\omega_n,$$

$$(7.48)$$

$$\hat{\xi} = \xi + \Delta\xi.$$

The variations of ω_n and ξ, denoted by $\Delta\omega_n$ and $\Delta\xi$, respectively, due to the inclusion of the parasitics of the active inductor are quantified by

$$\Delta\omega_n = \frac{1}{2}\frac{R_s}{R}\omega_n,$$

$$(7.49)$$

$$\Delta\xi = \frac{1}{4\xi}\frac{R_s}{R}.$$

It is seen that the loop bandwidth and damping factor are increased slightly. When C_p is considered, R is replaced with $R||\frac{1}{sC_p}$. Eq.(7.49) can still be used to quantify $\Delta\omega_n$ and $\Delta\xi$.

7.2.4 Phase Noise of Current-Mode PLLs

Consider the type I current-mode PLL shown in Fig.7.14 where the noise from the input, the noise of the active inductor loop filter, and the noise of the CCO are considered.

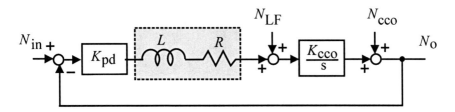

Figure 7.14. Phase noise analysis of type I current-mode PLLs with an ideal active inductor loop filter.

The transfer functions from these noise sources to the output of the PLL are given by

$$H_{in}(s) = \frac{N_o(s)}{N_{in}(s)} = \frac{\omega_n^2}{s^2 + 2\omega_n\xi s + \omega_n{}^2},$$

$$H_{LF}(s) = \frac{N_o(s)}{N_{LF}(s)} = \frac{K_{cco}\omega_{LF}\left(\frac{s}{\omega_{LF}} + 1\right)}{s^2 + 2\omega_n\xi s + \omega_n{}^2}, \qquad (7.50)$$

$$H_{cco}(s) = \frac{N_o(s)}{N_{cco}(s)} = \frac{\omega_{LF}s\left(\frac{s}{\omega_{LF}} + 1\right)}{s^2 + 2\omega_n\xi s + \omega_n{}^2}.$$

It is seen that the noise transfer function from the input to the output of the PLL have a low-pass characteristic with the cutoff frequency ω_n approximately and stopband attenuation of -40 dB/dec. Noise from the loop filter to the output of the PLL has a band-pass characteristic with passband center frequency ω_n. The noise from the CCO to the output of the PLL has a high-pass characteristic with the corner frequency ω_n. The Bodé plots of these transfer functions are shown in Fig.7.15.

Consider the type II current-mode PLL shown in Fig.7.16 where the noise from the input, the noise of the active inductor loop filter, and the noise of the CCO are considered. The transfer functions from these noise sources to the output of the PLL are given by

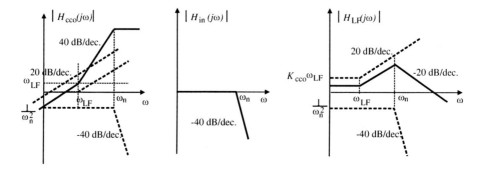

Figure 7.15. Bodé plot (magnitude) of the transfer function from noise sources to the output of type I current-mode PLLs.

$$H_{in}(s) = \frac{N_o(s)}{N_{in}(s)} = \frac{\omega_n^2\left(\dfrac{s}{\omega_{LF}} + 1\right)}{s^2 + 2\omega_n\xi s + \omega_n^2},$$

$$H_{LF}(s) = \frac{N_o(s)}{N_{LF}(s)} = \frac{K_{cco}s}{s^2 + 2\omega_n\xi s + \omega_n^2}, \qquad (7.51)$$

$$H_{cco}(s) = \frac{N_o(s)}{N_{cco}(s)} = \frac{s^2}{s^2 + 2\omega_n\xi s + \omega_n^2}.$$

It is seen that the noise transfer function from the input to the output of the PLL has a low-pass characteristic with the cut-off frequency ω_n and stopband attenuation rate -20 dB/dec. The noise from the CCO to the output of the PLL has a high pass characteristic with the corner frequency ω_n. The noise from the loop filter to the output of the PLL has a band-pass characteristic with passband center frequency ω_n.

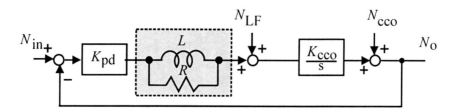

Figure 7.16. Phase noise analysis of type II current-mode PLLs with an ideal inductor loop filter.

7.2.5 Design Examples

This section compares the performance of two current-mode PLLs, one with an active inductor loop filter and an active inductor LC VCO and the other with an active inductor loop filter and a spiral LC VCO. Both operate at 3.0 GHz.

Current-Mode PLL with Active Inductor Loop Filter and Active Inductor LC VCO

The type I current-mode PLL proposed by DiClemente and Yuan in [66] consists of an active inductor loop filter and an active inductor CCO. The CCO of the PLL with its schematic shown in Fig.7.17 is a LC oscillator with Wu current-reuse active inductors. Frequency tuning of the CCO is achieved by changing the bias current J_2. J_1 is used to tune the resistance of the negative resistor so that the effect of the parasitic resistances of the active inductors is compensated for. The PLL was designed in TSMC-0.18μm 1.8V 6-metal 1-poly CMOS technology.

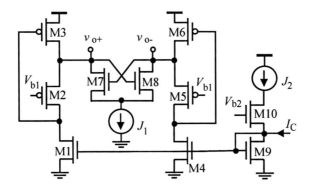

Figure 7.17. Simplified schematic of current-controlled oscillator with Wu current reuse active inductors.

The current-mode loop filter is constructed using differential-pair floating active inductor investigated in Chapter 2. The inductance of the active inductor is tuned by varying the tail current of the transconductors of the active inductor. The block diagram of the PLL is shown in Fig.7.18. The current-mode loop filter consists of a floating active inductor of an inductance 3 μH, in series with one 2.25 kΩ poly resistors on each side of the inductive inductor. The parasitic series resistance of the active inductor whose value is usually much smaller than 2.25 kΩ has a negligible impact on the loop dynamics of the PLL.

Fig.7.19 shows the dependence of the inductance of the active inductor on the dc biasing voltage V_{b2}. It is seen that an increase in the dc biasing voltage V_{b2} lowers the inductance. Also observed is that even when V_{b2} approaches zero, an inductive characteristic still exists. Moreover, the relation between

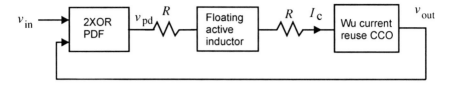

Figure 7.18.   Block diagram of the active inductor current-mode PLL by DiClemente and Yuan.

the inductance and the biasing voltage of the active inductor is approximately linear, enabling the smooth tuning of the inductance of the active inductor.

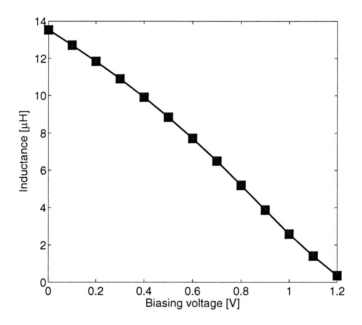

Figure 7.19.   Simulated dependence of the inductance of the floating active inductor on the biasing voltage V_{b2}. The inductance is evaluated at 3 GHz.

Fig.7.20 shows the dependence of the inductance of the active inductor on the dc voltage of the input of the active inductor. It is seen that even when the input transistors of the transconductors enter the triode region (high input dc voltage), the network still exhibits an inductive characteristic. The inductance of the active inductor in this case is larger due to the reduced transconductances of the transconductors. Also observed is that an inductive characteristic exists even when the input dc voltage is zero. The input transistors of the transconductors of the active inductor in this case are in its OFF state. The coupling between the input voltage and the branch current of the transconductors is due to the capacitances of the input transistors. The relation between the inductance and the swing of the input voltage of the active inductor is strongly nonlinear at both low and high input voltages.

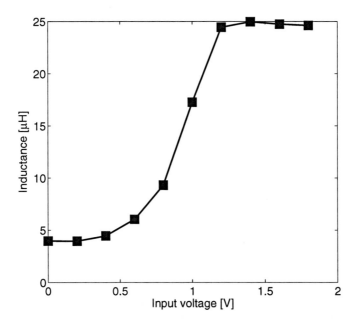

Figure 7.20. Simulated dependence of the inductance of Mahmoudi-Salama floating active inductor on the swing of the input voltage. The inductance is evaluated at 3 GHz.

Fig.7.21 shows the dependence of the output current of the current-mode loop filter on the duty cycle of the input voltage. It is evident that the dc component of the output current of the loop filter varies with the duty cycle of the input voltage.

It should be noted that although the input of the current-mode loop filter, which is the swing of the output voltage of the preceding phase detector, is typically rail-to-rail, the resistor preceding the active inductor and the gate capacitance of the input transistors of the active inductor form a low-pass RC network that limits the swing of the voltage at the gate of the input transistors of the active inductor to a rather small value. As a result, the input transistors of the transconductors of the active inductor remain in the saturation.

The phase detector of the PLL was a standard DFF phase/frequency detector where the DFF was implemented using TSPC logic to take the advantages of its simple configuration and low propagation delay [154]. Fig.7.22 plots the control current of the current-mode PLL. The phase noise of the PLL is plotted in Fig.7.23.

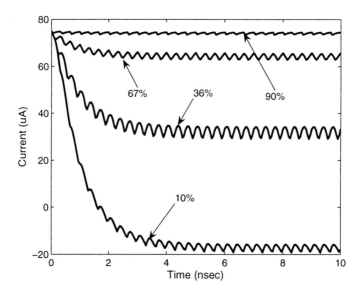

Figure 7.21.   Dependence of the output current of active inductor loop filter on the duty cycle of the input voltage (© IEEE 2007).

The power consumption of the phase detector, the current-mode loop filter, and the CCO were 0.09 mW, 0.1 mW, 12.2 mW, respectively. The power consumption of the current-mode loop filter is negligible, mainly due to its small dc biasing current required for obtaining a large inductance. The CCO dominated the power consumption of the PLL.

Current-Mode PLL with Active Inductor Loop Filter and Spiral Inductor LC VCO

It was pointed out that active inductors suffer from a high level of noise, as compared with their spiral counterparts. The simulation results of the preceding current-mode PLL with an active loop filter and an active inductor LC oscillator exhibits a high level of phase noise. Because active inductors were used in both the loop filter and the LC oscillator of the PLL, we are particularly interested in finding out whether the high level of the phase noise of the current-mode PLL is mainly caused by the noise of the active inductor loop filter or that of the active inductor LC oscillator. In the following current-mode PLL, we replace the active inductor LC oscillator with a spiral inductor LC oscillator. The loop filter is still active inductor based and the phase detector remains unchanged. Further, the replica-biasing technique is employed in the active inductor to minimize the effect of supply voltage fluctuation on the inductance of the active inductor. Fig.7.24 shows the schematic of the active inductor

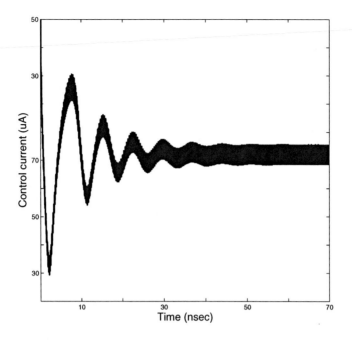

Figure 7.22. Simulated control current of the current-mode PLL with differential-pair floating active inductor loop filter and Wu current reuse active inductor CCO (© IEEE 2007).

of the loop filter with replica biasing [43, 44, 56]. The tail transistors of the transconductors are biased in the sub-threshold region to minimize g_{n1} and g_{m2} such that the inductance of the active inductor is maximized. This also minimizes the power consumption of the active inductor.

The schematic of the spiral inductor LC CCO is shown in Fig.7.25. The spiral inductors are replaced with the commonly used lumper model of spiral inductors in analysis of its performance [33]. Frequency tuning of the LC oscillator is achieved by varying the control voltage of two MOS varactors. The control voltage is controlled by a control current via an I/V converter. V_b tunes the resistance of the negative resistor.

Table 7.2 compares the effect of the fluctuation of the supply voltage on the inductance of the active inductor with and without the replica biasing. The inductance of the active inductor without the replica biasing is sensitive to V_{DD} fluctuation with a sensitivity of 0.167 nH/mV approximately. The sensitivity is reduced to 0.0235 nH/mV when the replica biasing is employed. Replica-biasing provides an economic and effective means to minimize the effect of V_{DD} fluctuation on the inductance of active inductors and should be used for all active inductors and transformers.

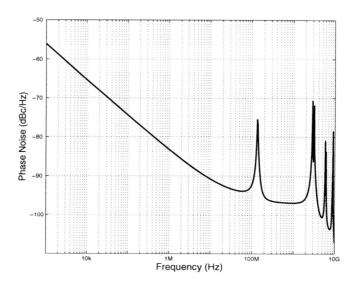

Figure 7.23. Simulated phase noise of the current-mode PLL with Mahmoudi-Salama floating active inductor loop filter and Wu current reuse active inductor CCO (© IEEE 2007).

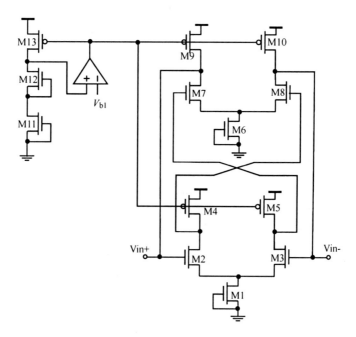

Figure 7.24. Simplified schematic of differential active inductor with replica biasing.

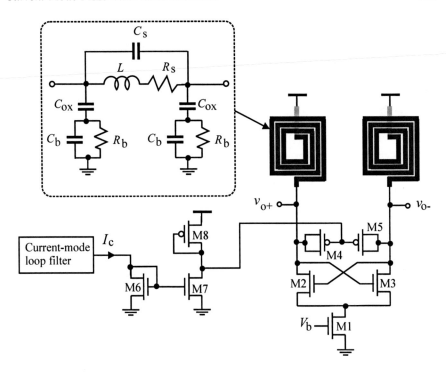

Figure 7.25. Simplified schematic of current-controlled LC oscillator with spiral inductors.

Table 7.2. Comparison of the effect of V_{DD} fluctuation on the inductance of the differential active inductor with and without replica biasing. The differences between the inductances of the active inductors with and without replica-biasing are due to the different dc biasing conditions of the active inductors.

V_{DD} [V]	Inductance [nH] (with replica biasing)	Inductance [nH] (without replica biasing)
1.70	193.2	154.2
1.74	193.7	146.6
1.78	192.8	139.6
1.82	191.4	133.4
1.86	189.8	126.9
1.90	188.5	120.7

The phase noise of the current-mode PLL is shown in Fig.7.26. It was obtained using Verilog-AMS based approach detailed earlier. The phase noise of the PLL with an active inductor loop filter and a passive LC oscillator is much lower as compared with that of the PLL with an active inductor loop filter and an active inductor LC oscillator investigated earlier. These results reveal that the phase noise of current-mode PLLs is dominated by that of the oscillator. The high level of the phase noise of the current-mode PLL with Wu active inductor LC oscillators is due to the poor phase noise performance of the oscillator. To minimize the phase noise of current-mode PLLs with active inductor loop filters and LC oscillators, minimizing of the phase noise of active inductor LC oscillators is the key.

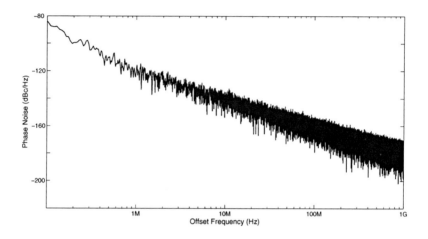

Figure 7.26.   Simulated phase noise of the current-mode PLL with the floating active inductor loop filter and the spiral inductor CCO of Fig.7.25.

The power consumption of the phase detector was approximately zero, the power consumption of the current-mode loop filter, and that of the CCO were 0.47 mW and 7.29 mW, respectively. The power consumption of the PLL is mainly due to that of the LC oscillator. The increased power consumption of the loop filter is due to the replica-biasing network.

7.3   Current-Mode PLLs with Active Transformers

This section explores the use of active transformers in current-mode phase-locked loops. Similar to current-mode PLLs with active inductors, a current-mode PLL with active transformers consists of a current-mode loop filter constructed using an active transformer with resistors, a current-controlled oscillator that is constructed using either active inductors or active transformers, and a phase detector, as shown in Fig.7.27. The current-mode loop filter is constructed using an active transformer with two primary windings and single secondary winding. As compared with current-mode loop filters with active inductors, the multiple primary windings of active transformers provide a convenient connection between the UP and DN output terminals of the preceding phase detector and the transformer. The large tunable inductances of the active transformer enables the loop filter to have a low cut-off frequency at the cost of a small silicon area.

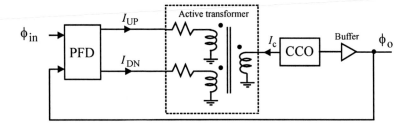

Figure 7.27. Configuration of active transformer current-mode PLLs.

7.3.1 Current-Mode Loop Filters with Active Transformers

A transformer-based current-mode low-pass filter can be constructed by connecting a resistor in series with the primary winding of a transformer, as shown in Fig.7.28. The output of the filter is the current of the secondary winding of the transformer.

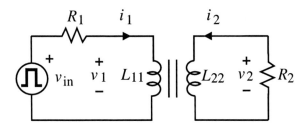

Figure 7.28. Configuration of current-mode transformer loop filters.

Assume that the transformer is ideal. Because for current-mode circuits, the load of the secondary winding is ideally zero, i.e. $R_2 = 0$, we have $V_2(s) = 0$. Further because

$$V_{in} = RI_1 + V_1,$$

$$V_1 = sL_{11}I_1 + sM_{12}I_2, \tag{7.52}$$

$$V_2 = sL_{22}I_2 + sM_{21}I_1,$$

where L_{11} and L_{22} are the self-inductance of the primary and secondary windings, respectively, M_{21} and M_{12} are the mutual inductance from the primary winding to the secondary winding and that from the secondary winding to the primary winding, respectively, we obtain the transfer function of the loop filter

$$\frac{I_L(s)}{V_{in}(s)} = \frac{M_{21}}{RL_{22}} \frac{1}{s\left(\dfrac{L_{11}L_{22} - M_{12}M_{21}}{RL_{22}}\right) + 1}. \tag{7.53}$$

It is evident from (7.53) that the filter is a low-pass with its cutoff frequency given by

$$\begin{aligned}
\omega_{-3dB} &= \frac{R}{L_{11}} \frac{1}{\left(1 - \dfrac{M_{12}M_{21}}{L_{11}L_{22}}\right)} \\
&= \frac{R}{L_{11}}\left(\frac{1}{1 - k_{12}k_{21}}\right).
\end{aligned} \tag{7.54}$$

Also note for a dc input ($s = 0$), the ac component of i_2 is zero as the primary and secondary windings are magnetically coupled. When i_2 is used to control a CCO in this case, the oscillation frequency of the CCO remains unchanged.

The simplified schematic of the current-mode loop filter is shown in Fig.7.29 for a nMOS active transformer and Fig.7.30 for a pMOS active transformer. The active transformers were proposed by Tang *et al.* in [149]. The coupling between the primary and secondary windings of the transformers is established using a series coupling mechanism. This differs from the parallel coupling mechanism detailed in Chapter 3. A notable drawback of the series-coupling is the large number of transistors stacked between V_{DD} and ground rails. The transformer is replica-biased to minimize the effect of V_{DD} fluctuation on the inductance of the active transformers, as per detailed in Chapter 3.

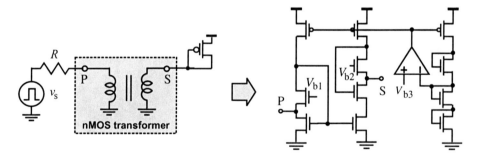

Figure 7.29.   Simplified schematic of current-mode active transformer loop filter (nMOS) with replica-biasing.

Fig.7.31 shows the dependence of the output current of the active transformer loop filter on the duty cycle of the input voltage. The dependence of the DC component of the output current on the duty cycle of the input voltage is evident.

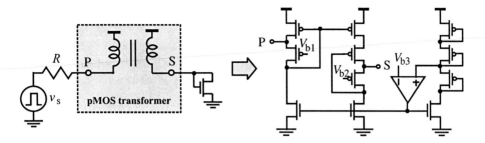

Figure 7.30. Simplified schematic of current-mode active transformer loop filter (pMOS) with replica-biasing.

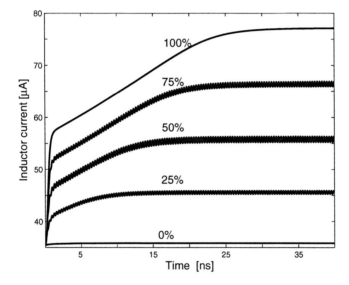

Figure 7.31. Simulated dependence of the output current of current-mode active transformer loop filter of Fig.7.29 on the duty cycle of the input voltage (© IEEE 2008).

Fig.7.32 plots the dependence of the bandwidth of nMOS / pMOS current-mode active transformer loop filter with and without replica-biasing. The loop filter with the replica-biased nMOS active transformer exhibits the lowest sensitivity to V_{DD}. The low sensitivity of the nMOS active transformer loop filter to V_{DD} fluctuation is mainly due to the fact that the nMOS active transformer is isolated from the supply voltage rail by three pMOS transistors that are biased in the saturation.

It should be noted that the bandwidth of the active transformer loop filter is also sensitive to process variation. Although the effect of process variation can be quantified in a statistical sense using either corner analysis or Monte

Carlo analysis, the exact amount of variation is not known in the design stage. The tunability of the cut-off frequency of the active transformer loop filter by varying the dc biasing currents of the active transformer allows users to adjust the inductances of the active transformer, subsequently the cut-off frequency of the loop filter, so that the desirable loop dynamics of current-mode PLLs can be obtained.

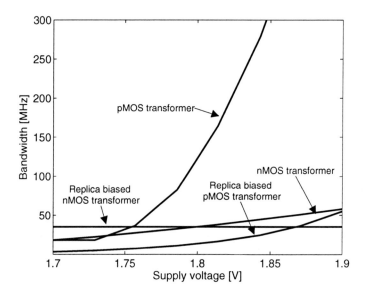

Figure 7.32.   Simulated sensitivity of the bandwidth of active-transformer loop filters to supply voltage (© IEEE 2008).

7.3.2   Loop Dynamics of Current-Mode PLLs

In this section, we investigate the loop dynamics of current-mode active transformer PLLs in the lock state. To simplify analysis, the transformer of the loop filter is assumed to be ideal first. As shown in Fig.7.33, a resistor is in series with the transformer to form the loop filter. The voltage at the input of the loop filter is given by

$$V_{pd} = RI_1 + sL_{11}I_1 + sM_{12}I_2. \tag{7.55}$$

Because the active transformer is uni-directional, $M_{12} = 0$. For the secondary winding,

$$V_2 = sL_{22}I_2 + sM_{21}I_1. \tag{7.56}$$

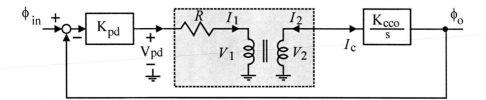

Figure 7.33.   Analysis of loop dynamics of transformer current-mode loop filter.

Further because the load of the secondary winding is a CCO whose input impedance is zero ideally, we have $V_2 = 0$. As a result,

$$I_c = \frac{M_{21}}{L_{22}(sL_{11} + R)} V_{pd}. \tag{7.57}$$

Eq.(7.57) shows that the loop filter is a low-pass with its bandwidth given by

$$\omega_{LF} = \frac{R}{L_{11}}. \tag{7.58}$$

The loop gain of the PLL is given by

$$H_o(s) = K_{pd} \left[\frac{M_{21}}{L_{22}(sL_{11} + R)} \right] \left(\frac{K_{cco}}{s} \right). \tag{7.59}$$

The PLL is type I as the loop gain only has one pole at the origin. The closed-loop transfer function of the PLL can also be derived.

$$H_c(s) = \frac{\Phi_o(s)}{\Phi_i(s)} = \frac{K_{pd} K_{cco} M_{21}}{L_{11} L_{22}} \frac{1}{s^2 + s\dfrac{R}{L_{11}} + \dfrac{K_{pd} K_{cco} M_{21}}{L_{11} L_{22}}}. \tag{7.60}$$

The loop bandwidth ω_n and damping factor ς of the PLL are obtained from (7.60)

$$\omega_n = \sqrt{\frac{K_{pd} K_{cco} M_{21}}{L_{11} L_{22}}},$$

$$\xi = \frac{R}{2} \sqrt{\frac{L_{22}}{L_{11} K_{pd} K_{cco} M_{21}}}. \tag{7.61}$$

Also

$$\omega_n \xi = \frac{\omega_{LF}}{2}. \tag{7.62}$$

The poles of $H_c(s)$ are given by

$$s_{1,2} = \frac{\omega_{LF}}{2}\left(-1 \pm \sqrt{1 - \frac{1}{\xi^2}}\right). \tag{7.63}$$

We examine the preceding results in detail :

- The damping factor ξ can be tuned by varying R without affecting the loop bandwidth. This property is very similar to that of type I active inductor current-mode PLLs, and is different from that of type I voltage-mode PLLs.

- The higher the self inductance of the primary winding L_{11}, the smaller the loop bandwidth ω_n and the smaller the damping factor ξ. As a result, the attenuation of the amplitude of the oscillating response of the PLL is reduced. This is echoed with a longer lock time.

- The higher the self inductance of the secondary winding L_{22}, the smaller the loop bandwidth ω_n and the larger the damping factor ξ. The product, $\omega_n\xi$, however, remains unchanged. As a result, the attenuation of the amplitude of the oscillating response of the PLL remains unchanged.

- The larger the mutual inductance M_{21}, the larger the loop bandwidth and the smaller the damping factor. Because $\omega_n\xi$ remains unchanged, the attenuation of the amplitude of the oscillating output of the PLL is the same. The oscillating frequency, however, will increase.

- The larger the resistance R, the larger the damping factor ξ, the larger the attenuation of the amplitude of the oscillating output of the PLL.

The preceding analysis is based on the assumption that the transformer of the current-mode loop filter is ideal. In what follows we include the parasitic resistances of active transformers and investigate the effect of these parasitic resistances on the performance of the loop filter. Because the active transformer used in the loop filter is uni-directional, we have

$$V_{pd} = RI_1 + (sL_{11} + R_{11})I_1. \tag{7.64}$$

Further because

$$V_2 = (sL_{22} + R_{22})I_2 + (sM_{21} + R_{21})I_1, \tag{7.65}$$

and $V_2 = 0$, we arrive at

$$H_{LF}(s) = \frac{I_c(s)}{V_{pd}(s)} = \frac{sM_{21} + R_{21}}{(sL_{22} + R_{22})(sL_{11} + R_{11} + R)}. \tag{7.66}$$

The PLL is still type I as there is only one pole at the origin. The closed-loop transfer function of the PLL is given by

$$H_c(s) = \frac{K_{pd}K_{cco}(sM_{21} + R_{21})}{s(sL_{22} + R_{22})(sL_{11} + R_{11}) + K_{pd}K_{cco}(sM_{21} + R_{21})}. \tag{7.67}$$

It is readily to verify that if the transformer is ideal, i.e. $R_{11}, R_{22}, R_{21} = 0$, Eq.(7.67) is simplified to (7.60). To simplify (7.67), we notice that the coupling between the two winding inductors of the active transformer is via transconductors, typically MOS transistors, as shown in Chapter 3, there is no direct current path between the two active inductors. As a result, $R_{21} = R_{12} = 0$. Eq.(7.67) becomes

$$H_c(s) = \left(\frac{K_{pd}K_{cco}M_{21}}{L_{11}L_{22}}\right) \frac{1}{s^2 + s\left(\frac{R_{11} + R}{L_{11} + \frac{R_{22}}{L_{22}}}\right) + \frac{K_{pd}K_{cco}M_{21} + R_{22}(R_{11} + R)}{L_{11}L_{22}}}$$

$$\tag{7.68}$$

The loop bandwidth and the damping factor are obtained from (7.68)

$$\hat{\omega}_n = \omega_n\sqrt{1 + \frac{R_{22}(R_{11} + R)}{K_{pd}K_{cco}M_{21}}}, \tag{7.69}$$

$$\hat{\xi} = \hat{\xi}_1 + \hat{\xi}_2,$$

where

$$\hat{\xi}_1 = \xi\left(1 + \frac{R_{11}}{R}\right)\sqrt{\frac{1}{1 + \frac{R_{22}(R_{11} + R)}{K_{pd}K_{cco}M_{21}}}},$$

$$\tag{7.70}$$

$$\hat{\xi}_2 = \xi\left(\frac{R_{22}}{2}\frac{L_{11}}{L_{22}}\right)\sqrt{\frac{1}{1 + \frac{R_{22}(R_{11} + R)}{K_{pd}K_{cco}M_{21}}}},$$

Making use of the first-order approximation,

$$\sqrt{\frac{1}{1+x}} \approx 1 - \frac{1}{2}x, \qquad (x < 1), \qquad (7.71)$$

we arrive at

$$\hat{\omega}_n \approx \omega_n \left[1 - \frac{1}{2} \frac{R_{22}(R_{11} + R)}{K_{pd}K_{cco}M_{21}} \right], \qquad (7.72)$$

$$\hat{\xi} = \hat{\xi}_1 + \hat{\xi}_2,$$

where

$$\hat{\xi}_1 \approx \xi \left(1 + \frac{R_{11}}{R} \right) \left[1 - \frac{1}{2} \frac{R_{22}(R_{11} + R)}{K_{pd}K_{cco}M_{21}} \right],$$

$$\hat{\xi}_2 = \xi \left(\frac{R_{22}}{2} \frac{L_{11}}{L_{22}} \right) \left[1 - \frac{1}{2} \frac{R_{22}(R_{11} + R)}{K_{pd}K_{cco}M_{21}} \right], \qquad (7.73)$$

7.3.3   Phase Noise of Current-Mode PLLs

The phase noise of current-mode PLLs with active transformers in the lock state can be analyzed in a similar way as that of current-mode PLLs with active inductors. Consider Fig.7.34 where the noise from the input, the noise of the loop filter, and the noise of the CCO are considered. The phase noise at the output of the PLL is given by

$$H_{cco}(s) = \frac{N_o(s)}{N_{cco}(s)} = \left(\frac{R}{L_{11}} \right) \frac{s\left(\dfrac{s}{\omega_{LF}} + 1 \right)}{s^2 + 2\omega_n \xi s + \omega_n^2},$$

$$H_{LF}(s) = \frac{N_o(s)}{N_{LF}(s)} = \left(\frac{K_{cco}R}{L_{11}} \right) \frac{\dfrac{s}{\omega_{LF}} + 1}{s^2 + 2\omega_n \xi s + \omega_n^2}, \qquad (7.74)$$

$$H_{in}(s) = \frac{N_o(s)}{N_{in}(s)} = \frac{\omega_n^2}{s^2 + 2\omega_n \xi s + \omega_n^2}.$$

It is seen that the noise from the input to the output of the PLL has a low-pass characteristic with the cutoff frequency ω_n and stopband attenuation -40 dB/dec.

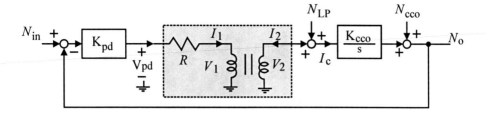

Figure 7.34. Noise analysis of current-mode PLLs with an ideal transformer loop filter.

The noise from the loop filter to the output of the PLL has a low-pass characteristic with the cut-off frequency ω_n. The attenuation in the stopband is only -20 dB/dec. The noise from the CCO to the output of the PLL has a high-pass characteristic with the corner frequency ω_n. Fig.7.35 shows the Bodé plot (magnitude) of the transfer functions from the noise sources to the output of the active transformer PLLs.

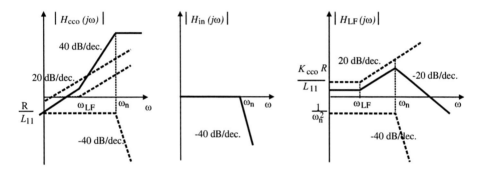

Figure 7.35. Bodé plot (magnitude) of the transfer functions from the noise sources to the output of active transformer PLLs.

7.3.4 Design Example

This section presents the design of a 3 GHz active transformer PLL implemented in TSMC-0.18μm 1.8V CMOS technology [155]. The active inductor CCO is a 3 GHz LC oscillator with Lu active inductors investigated in Chapter 2. The simplified schematic of the CCO is shown in Fig.7.36. The reported phase noise measurement of the oscillator in [60] was from -101 dBc/Hz to -118 dBc/Hz at 1 MHz frequency offset. As pointed out earlier in current-mode PLLs with active inductors that the phase noise of these PLLs is dominated by that of the LC oscillators. The simulated phase noise of the CCO in Fig.7.36 is plotted in Fig.7.37. The simulation results agree with the published measurement results, demonstrating that the LC oscillator with Lu active inductors offers a lower level of phase noise, as compared with the LC oscillator with Wu current-reuse active inductors investigated earlier.

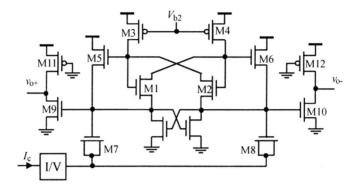

Figure 7.36.   Simplified schematic of current-mode LC oscillator with Lu active inductors.

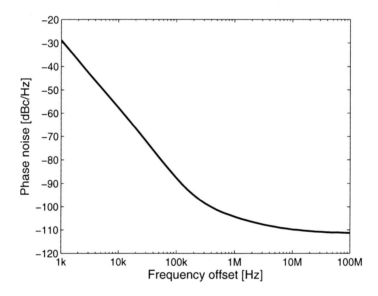

Figure 7.37.   Simulated phase noise of current-mode LC oscillator with Lu active inductors.

Fig.7.38 plots the control current of the PLL. The phase noise of the PLL was analyzed using the Kundert's Verilog-AMS approach detailed earlier and the results are shown in Fig.7.39. The phase noise of the PLL is below -100 dBc/Hz at 1 MHz frequency offset. The power consumption of the CCO, the loop filter, and the PFD was 15 mW, 0.138 mW, and 0.864 mW, respectively.

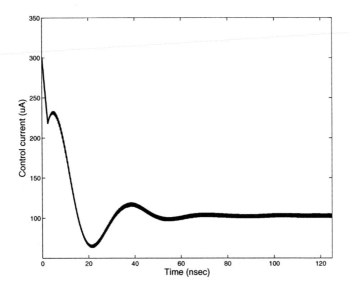

Figure 7.38. Simulated control current of the active transformer current-mode PLL with the active transformer loop filter and Lu active inductor CCO (© IEEE 2008).

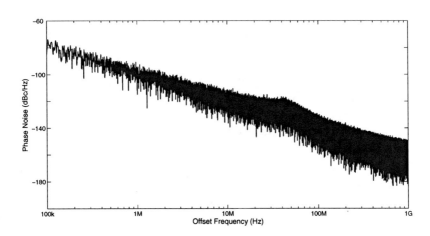

Figure 7.39. Simulated phase noise of the active transformer current-mode PLL with the active transformer loop filter and Lu active inductor CCO (© IEEE 2008).

7.4 Chapter Summary

The fundamentals of voltage-mode PLLs including classifications, loop dynamics, and phase noise have been presented. We have shown that PLLs can be categorized into type I and type II, depending upon the number of poles of the loop gain of the PLLs at the origin. High-order PLLs are rarely used due to their poor stability. Type I voltage-mode PLLs suffer from the main drawback that the loop bandwidth and damping factor can not be tuned independently. Type II voltage-mode PLLs, on the other hand, can adjust their loop bandwidth and damping factor individually, and are the most widely used PLLs. The phase noise of type II voltage-mode PLLs in the lock state has also been analyzed. The transfer characteristic from the noise source of the oscillator to the output of a type II PLL is high-pass with the corner frequency ω_h, the loop bandwidth. The noise of the loop filter to the output of the PLL has a band-pass characteristic with its center frequency ω_n. The transfer characteristic from the noise at the input of the PLL to the output of the PLL is low-pass with its cutoff frequency ω_n and stopband attenuation -40 dB/dec. The noise from the VCO contributes the most to the phase noise of the PLL at frequencies beyond ω_h, the noise from the input of the PLL contributes the most to the phase noise of the PLL at frequencies below ω_n and the noise from the loop filter contributes the most to the phase noise of the PLL in the vicinity of ω_h.

The configurations and characteristics of current-mode PLLs with active inductors have been presented. We have shown that both type I and type II current-mode PLLs can be constructed by employing active inductor loop filters. These current-mode filters sustain the output current by employing inductors. The dc component of the output current of the current-mode loop filter is proportional to the duty cycle of the input voltage. Unlike type I voltage-mode PLLs, the loop bandwidth and damping factor of both types I and II current-mode PLLs can be tuned individually. The use of active inductors in current-mode loop filters not only significantly reduces the silicon area needed for fabricating the loop filter, the tunability of the inductance of active inductors also enables the tuning of the loop dynamics of current-mode PLLs so that not only desired loop dynamics can be obtained, the effect of process variation can also be compensated for effectively. Phase noise analysis of types I and II current-mode PLLs have also been presented. Trade-offs between loop bandwidth and the phase noise of current-mode PLLs are similar to those of voltage-mode PLLs. Simulation results have demonstrated that the power consumption and noise contribution of active inductor loop filters are negligible. The phase noise and power consumption of current-mode PLLs with an active inductor loop filter and an active inductor LC current-controlled oscillator are dominated by those of the CCO.

The configurations and characteristics of current-mode PLLs with active transformers have also been investigated in detail. We have shown that current-mode loop filters can be constructed by using active transformers of two primary windings and one secondary winding. The dc component of the output current of the secondary winding of the active transformer is proportional to the duty cycle of the input voltage at the primary winding. Both the loop bandwidth and damping factor of current-mode PLLs with active transformers can be tuned individually. The use of active transformers in current-mode loop filters not only significantly reduces the silicon area needed for fabricating the loop filter, the tunability of the inductances of active transformers also enables the tuning of the loop dynamics of current-mode PLLs so that the effect of process variation can be compensated for effectively. Phase noise analysis of current-mode PLLs with active transformers have also been conducted. Trade-offs between the loop bandwidth and phase noise of current-mode PLLs with active transformers have been derived. Again, the phase noise and power consumption of current-mode PLLs with an active transformer loop filter and an active inductor LC current-controlled oscillator are dominated by those of the CCO.

To minimize the effect of supply voltage fluctuation on the performance of both current-mode loop filters and current-controlled oscillators with active inductors or active transformers, replica-biasing is mandatory.

References

[1] M. Straayer, J. Cabanillas, and G. Rebeiz. "A low-noise transformer-based 1.7 GHz CMOS VCO". In *Proc. IEEE Int'l Conf. Solid-State Circuits*, volume 1, pages 286–287, Feb. 2002.

[2] D. Baek, T. Song, E. Yoon, and S. Hong. "8-GHz CMOS quadrature VCO using transformer-based LC tank". *IEEE Microwave and Wireless Component Letters*, 13(10):446–448, Oct. 2003.

[3] H. Wang, M. Lin, Y. Li, and H. Chen. "Some design aspects on 5 GHz CMOS quadrature VCO with fully integrated LC-tank". *Proc. 5th Int'l Conf. ASIC*, 2:1010–1013, Oct. 2003.

[4] S. Gierkink, S. Levantino, R. Frye, C. Samori, and V. Boccuzzi. "A low-phase-noise 5-GHz CMOS quadrature VCO using superharmonic coupling". *IEEE J. Solid-State Circuits*, 38(7):1148–1154, Jul. 2003.

[5] H. Choi, S. Shin, and S. Lee. "A low-phase noise LC-QVCO in CMOS technology". *IEEE Microwave and Wireless Component Letters*, 14(11):540–542, Nov. 2004.

[6] R. Dehghani and S. Atarodi. "Design of an optimized 2.5 GHz CMOS differential LC oscillator". *IEE Proceedings - Microwave, Antenna, and Propagation*, 151(2):167–172, April 2004.

[7] J. Kim, Y. Lee, and S. Park. "Low-noise CMOS LC oscillator with dual-ring structure". *IEE Electronics Letters*, 40(17):1031–1032, Aug. 2004.

[8] J. Chang and C. Kim. "A symmetrical 6-GHz fully integrated cascode coupling CMOS LC quadrature VCO". *IEEE Microwave and Wireless Components Letters*, 15(10):670–672, Oct. 2005.

[9] N. Oh and S. Lee. "11-GHz CMOS differential VCO with back-gate transfer feedback". *IEEE Microwave and Wireless Component Letters*, 15(11):733–735, Nov. 2005.

[10] B. Soltanian and P. Kinget. "Tail current-shaping to improve phase noise in LC voltage-controlled oscillators". *IEEE J. Solid-State Circuits*, 41(8):1792–1802, Aug. 2006.

[11] S. Mohan, S. Hershenson, M. Boyd, and T. Lee. "Simple accurate expressions for planar spiral inductances". *IEEE J. Solid-State Circuits*, 34(10):1419–1424, Oct. 1999.

[12] S. Mohan, S. Hershenson, M. Boyd, and T. Lee. "Bandwidth extension in CMOS with optimized on-chip inductors". *IEEE J. Solid-State Circuits*, 35(3):346–355, Mar. 2000.

[13] B. Sun and F. Yuan. "A new inductor series-peaking technique for bandwidth enhancement of CMOS current-mode circuits". *Analog Integrated Circuits and Signal Processing*, 37(3):259–264, Dec.. 2003.

[14] F. Yuan. *CMOS current-mode circuits for data communications*. Springer, New York, 2007.

[15] B. Ballweber, R. Gupta, and D. Allstot. "A fully integrated 0.5-5.5 GHz CMOS distributed amplifier'. *IEEE J. Solid-State Circuits*, 35(2):231–239, Jan. 2000.

[16] H. Ahn and D. Allstot. "A 0.5-8.5 GHz fully differential CMOS distributed amplifier". *IEEE J. Solid-State Circuits*, 37(8):985–993, Aug. 2002.

[17] B. Razavi. *RF microelectronics*. Prentice-Hall, Upper Saddle River, N.J., 1998.

[18] B. Leung. *VLSI for wireless communication*. Prentice-Hall, Upper Saddle River, NJ, 2002.

[19] M. Tsai and H. Wang. "A 0.3-2.5 GHz ultra-wideband mixer using commercial 0.18μm CMOS technology". *IEEE Microwave and Wireless Component Letters*, 14(11):522–524, Nov. 2004.

[20] L. Lu and Y. Liao. "A 4-GHz phase shifter MMIC in 0.18-μm CMOS". *IEEE Microwave and Wireless Components Letters*, 15(10):694–696, Oct. 2005.

[21] W. Kuhn, F. Stephenson, and A. Elshabini-Riad. "A 200 MHz CMOS Q-enhanced LC bandpass filter". *IEEE J. Solid-State Circuits*, 31(8):1112–1122, Aug. 1996.

[22] R. Duncan, K. Martin, and A. Sedra. "A Q-enhanced active-RLC bandpass filter". *IEEE Trans. Circuits and Systems II*, 44(5):341–347, May 1997.

[23] T. Soorapanth and S. Wong. "A 0-dB IL 2140±30 MHz bandpass filter utilizing Q-enhanced spiral inductors in standard CMOS". *IEEE J. Solid-State Circuits*, 37(5):579–586, May 2002.

[24] F. Dulger, E. Sanchez-Sinencio, and J. Silva-Martinez. "A 1.3-V 5-mW fully integrated tunable bandpass filter at 2.1 GHz in 0.35μm CMOS". *IEEE J. Solid-State Circuits*, 38(6):918–928, June 2003.

[25] W. Kuhn, D. Nodde, D. Kelly, and A. Orsborn. "Dynamic range performance of on-chip RF bandpass filters". *IEEE Trans. on Circuits and Systems II.*, 50(10):685–694, Oct. 2003.

[26] S. Bantas and Y. Koutsoyannopoulos. "CMOS active-LC bandpass filters with coupled-inductor Q-enhancement and center frequency tuning". *IEEE J. Solid-State Circuits*, 51(2):69–76, Feb. 2004.

[27] J. Kulyk and J. Haslett. "A monolithic CMOS 2368±30 MHz transformer based Q-enhanced series-C coupled resonator bandpass filter". *IEEE J. Solid-State Circuits*, 41(2):362–374, Feb. 2006.

[28] L. Lu, Y. Liao, and C. Wu. "A miniaturized Wilkinsonpower divider with CMOS ac inductors". *IEEE Microwave Wireless Components Letters*, 15(11):775–777, Nov. 2005.

[29] C. Yue and S. Wong. "On-chip spiral inductors with patterned ground shields for Si-based RF IC". *IEEE J. Solid-State Circuits*, 33(5):743–752, May 1998.

[30] A. Zolfaghari, A. Chan, and B. Razavi. "Stacked inductors and transformers in CMOS technology". *IEEE J. Solid-State Circuits*, 36(4):620–628, Apr. 2001.

[31] Y. Lin, C. Chen, H. Liang, and C. Chen. "High-performance on-chip transformers with partial polysilicon patterned ground shields (PGS)". *IEEE Trans. Electron Devices*, 54(1):157–160, Jan. 2007.

[32] B. Owens, S. Adluri, P. Birrir, R. Shreeve, K. Mayaram S. Arunachalam, and T. Fiez. "Simulation and measurement of supply and substrate noise in mixed-signal ICs". *IEEE J. Solid-State Circuits*, 40(2):382–391, Feb. 2005.

[33] C. Yue, C. Ryu, J. Lau, T. Lee, and S. Wong. "A physical model for planar spiral inductor on silicon". In *Proc. Int'l Electron Devices Meeting*, pages 155–158, Dec. 1996.

[34] O. Kenneth. "Estimation methods for quality factors of inductors fabricated in silicon integrated circuit process technologies". *IEEE J. Solid-State Circuits*, 33(8):1249–1252, Aug. 1998.

[35] Y. Cao, R. Groves, X. Huang, N. Zamdmer, J. Plouchart, R. Wachnik, T. King, and C. Hu. "Frequency-dependent equivalent-circuit model for on-chip spiral inductors". *IEEE J. Solid-State Circuits*, 38(3):419–426, Mar. 2003.

[36] Y. Wu, X. Ding, M. Ismail, and H. Olsson. "RF band-pass filter design based on CMOS active inductors". *IEEE Trans. Circuits and Systems II*, 50(12):942–949, Dec. 2003.

[37] A. Thanachayanont and A. Payne. "CMOS floating active inductor and its applications to band-pass filter and oscillator design". *IEE Proceedings, Part G - Circuits, Devices, and Systems*, 147(1):42–48, Feb. 2000.

[38] A. Thanachayanont. "A 1.5-V high-Q CMOS active inductor for IF/RF wireless applications". In *Proc. IEEE Asia-Pacific Conf. Circuits Syst.*, volume 1, pages 654–657, 2000.

[39] T. Lin and A. Payne. "Design of a low-voltage, low-power, wide-tuning integrated oscillator". In *Proc. IEEE Int'l Symp. Circuits Syst.*, volume 5, pages 629–632, Geneva, Switzerland, May 2000.

[40] E. Sackinger and W. Fischer. "A 3 GHz 32 dB CMOS limiting amplifier for SONET OC-48 receivers". In *Proc. Int'l Solid-State Circuits Conf.*, volume Proc. IEEE Int'l Solid-State Circuit Conf., page 158, Feb. 2000.

[41] E. Sackinger and W. Fischer. "A 3-GHz 32-dB CMOS limiting amplifier for SONET OC-48 receivers". *IEEE J. Solid-State Circuits*, 35(12):1884–1888, Dec. 2000.

[42] Y. Wu, M. Ismail, and H. Olsson. "CMOS VHF/RF CCO based on active inductors". *IEE Electronics Letters*, 37(8):472–473, Apr. 2001.

[43] M. Grozing, A. Pascht, and M. Berroth. "A 2.5 V CMOS differential active inductor with tunable L and Q for frequencies up to 5 GHz". In *Proc. Int'l Microwave Symp.*, volume 1, pages 575–578, Phoenix, May 2001.

[44] M. Grozing, A. Pascht, and M. Berroth. "A 2.5 V CMOS differential active inductor with tunable L and Q for frequencies up to 5 GHz". In *Proc. IEEE Radio Freq. Integrated Circuits Symp.*, pages 271–274, 2001.

[45] Y. Wu, X. Ding, M. Ismail, and H. Olsson. "Inductor-less CMOS RF band-pass filter". *IEE Electronics Letters*, 37(16):1027–1028, Aug. 2001.

[46] Y. Wu, C. Shi, X. Ding, M. Ismail, and H. Olsson. "Design of CMOS VHF/RF biquadratic filters". *Analog Integrated Circuits and Signal Processing*, 33:239–248, 2002.

[47] A. Thanachayanont. "Low-voltage low-power high-Q CMOS RF bandpass filter". *IEE Electronics Letters*, 38(13):615–616, June 2002.

[48] H. Xiao and R. Schaumann. "Very-high-frequency low-pass filter based on a CMOS active inductor". In *Proc. IEEE Int'l Symp. Circuits Syst.*, volume 2, pages 1–4, May 2002.

[49] H. Xiao, R. Schaumann, and W. Daasch. "A radio-frequency CMOS active inductor and its application in designing high-Q filters". In *Proc. IEEE Int'l Symp. Circuits Syst.*, volume 4, pages 197–200, Vancouver, May 2004.

[50] J. Liang, C. Ho, C. Kuo, and Y. Chan. "CMOS RF band-pass filter design using the high quality active inductor". *IEICE Trans. on Electron*, E88-C(12):2372–2376, Dec. 2005.

[51] Z. Gao, M. Yu, Y. Ye, and J. Ma. "Wide tuning range of a CMOS RF bandpass filter for wireless applications". In *Proc. IEEE Conf. Electron Devices & Solid-State Circuits*, pages 53–56, Dec. 2005.

[52] Z. Gao, M. Yu, Y. Ye, and J. Ma. "A CMOS RF tuning wide-band bandpass filter for wireless applications". In *Proc. IEEE Int'l Conf. SOC.*, pages 79–80, Spet. 2005.

[53] Z. Gao, M. Yu, Y. Ye, and J. Ma. "A CMOS RF bandpass filter based on the active inductor". In *Proc. Int'l Conf. ASIC.*, pages 604–607, Oct. 2005.

[54] J. Chen, G. Sheets, C. Guo, F. Saibi, F. Yang, K. Azadet, J. Lin, and G. Zhang. "Electrical backplane equalization using programmable analog zeros and folded active inductors". *IEEE Tran. on Microwave Theory Tech.*, 55(7):1366–1369, July 2005.

[55] J. Chen, G. Sheets, C. Guo, F. Saibi, F. Yang, K. Azadet, J. Lin, and G. Zhang. "Electrical backplane equalization using programmable analog zeros and folded active inductors". In *Proc. IEEE MidWest Symp. Circuits Syst.*, pages 1366–1369, Aug. 2005.

[56] F. Mahmoudi and C. Salama. "8 GHz tunable CMOS quadrature generator using differential active inductors". In *Proc. IEEE Int'l Symp. Circuits Syst.*, volume 3, pages 2112–2115, May 2005.

[57] J. Jiang and F. Yuan. "A new CMOS current-mode multiplexer for 10Gbps serial links". *Analog Integrated Circuits and Signal Processing*, 44(1):61–67, Jul. 2005.

[58] F. Yuan. "A modified Park-Kim voltage-controlled ring oscillator for multi-Gbps serial links". *Analog Integrated Circuits and Signal Processing*, 47(3), June 2006.

[59] F. Yuan. "A fully differential VCO cell with active inductors for Gbps serial links". *Analog Integrated Circuits and Signal Processing*, 47(2), May 2006.

[60] L. Lu, H. Hsieh, and Y. Liao. "A wide tuning-range CMOS VCO with a differential tunable active inductor". *IEEE Trans. on Microwave Theory Appl.*, 54(9):3462–3468, Sept. 2006.

[61] H. Xiao and R. Schaumann. "A 5.4-Ghz high-Q tunable active-inductor bandpass filter in standard digital CMOS technology". *Analog Integrated Circuits and Signal Processing*, 51(1):1–9, Apr. 2007.

[62] R. Weng and R. Kuo. "An ω_o-Q tunable CMOS active inductor for RF bandpass filters". In *Proc. Int'l Symp. Signals, Systems, and Electronics*, pages 571–574, Aug. 2007.

[63] A. Tang, F. Yuan, and E. Law. "A new CMOS active transformer QPSK modulator with optimal bandwidth control". *IEEE Trans. on Circuits Syst. II.*, 55(1):11–15, Jan. 2008.

[64] A. Tang, F. Yuan, and E. Law. "CMOS class AB active transformers with applications in LC oscillators". In *IEEE Int'l Symp. Signals, Systems and Electronics*, pages 501–504, Montreal, Aug. 2007.

[65] A. Thanachayanont and S. Ngow. "Class AB VHF CMOS active inductor". In *Proc. IEEE Mid-West Symp. Circuits Syst.*, volume 1, pages 64–67, Aug. 2002.

[66] D. DiClemente and F. Yuan. "Current-mode phase-locked loops : a new architecture". *IEEE Trans. on Circuits Syst. II.*, 54(4):303–307, Apr. 2007.

[67] A. Tang, F. Yuan, and E. Law. "A new constant-Q active inductor with applications in low-noise oscillator". *IEE Electronics Letters*, Oct. 2007 (submitted).

[68] B. Razavi. *Design of CMOS integrated circuits for optical communications*. McGraw-Hill, Boston, 2003.

[69] B. Razavi. "A study of phase noise in CMOS oscillators". *IEEE J. Solid-State Circuits*, 31(3):331–343, Mar. 1996.

[70] B. Razavi. *Design of analog CMOS integrated circuits*. McGraw-Hill, Boston, 2001.

[71] P. Gray, P. Hurst, S. Lewis, and R. Meyer. *Analysis and design of analog integrated circuits*. John Wiley and Sons, New York, 4th edition, 2001.

[72] J. Vlach and K. Singhal. *Computer methods for circuit analysis and design*. Van Nostrand Reinhold, New York, 2nd edition, 1994.

[73] A. Thanachayanont. "CMOS transistor-only active inductor or IF/RF applications". In *Proc. IEEE Int'l Industrial Tech. Conf.*, volume 2, pages 1209–1212, Bangkok, 2002.

[74] A. Karsilayan and R. Schaumann. "A high-frequency high-Q CMOS active inductor with DC bias control". In *Proc. IEEE Mid-West Symp. Circuits Syst.*, pages 486–489, Lansing, Aug. 2000.

[75] H. Xiao and R. Schaumann. "A low-voltage low-power CMOS 5GHz oscillator based on active inductors". In *Proc. IEEE Int'l Symp. Circuits Syst.*, volume 1, pages 231–234, Sept. 2002.

[76] S. Ngow and A. Thanachayanont. "A low-voltage wide dynamic range CMOS floating active inductor". In *Proc. Conf. Convergent Technologies for Asia-Pacific Region*, volume 4, pages 1640–1643, Oct. 2003.

[77] S. Hara, T. Tokumitsu, T. Tanaka, and M. Aikawa. "Broadband monolithic microwave active inductor and its application to miniaturized wideband amplifiers". *IEEE Trans. Microwave Theory and Applications*, 36(12):1920–1924, Dec. 1988.

[78] S. Hara, T. Tokumitsu, T. Tanaka, and M. Aikawa. "Lossless broad-band monolithic microwave active inductors". *IEEE Trans. Microwave Theory and Applications*, 37(12):1979–1984, Dec. 1989.

[79] W. Chen and C. Lu. "A 2.5 Gbps CMOS optical receiver analog front-end". In *Proc. IEEE Custom Integrated Circuits Conf.*, pages 359–362, May 2002.

[80] G. Chen, W. Chen, and R. Luo. "A 2.5 Gbps CMOS laser diode driver with preemphasis technique". In *Proc. IEEE Asia-Pacific Conf. Circuits Syst.*, pages 65–68, Aug. 2002.

[81] S. Song, S. Park, and H. Yoo. "A 4-Gb/s CMOS clock and data recovery circuit using 1/8-rate clock technique". *IEEE J. Solid-State Circuits*, 38(7):1213–1219, Jul. 2003.

[82] Y. Wang, M. Khan, S. Ali, and R. Raut. "A fully differential CMOS limiting amplifier with active inductor for optical receiver". In *Proc. IEEE Canadian Conf. Elec. Comp. Eng.*, pages 1751–1754, 2005.

[83] C. Wu, H. Liao, and S. Liu. "A 1 V 4.2 mW fully integrated 2.5 Gb/s CMOS limiting amplifier using folded active inductors". In *Proc. Int'l Symp. Circuits Syst.*, volume 1, pages 1044–1047, May 2004.

[84] A. Thanachayanont. "CMOS transistor-only active inductor for IF/RF applications". In *Proc. ICIT*, pages 1209–1212, Bangkok, 2002.

[85] U. Yodprasit and J. Ngarmnil. "Q-enhanced technique for RF CMOS active inductor". In *Proc. IEEE Int'l Symp. Circuits Syst.*, volume 5, pages 589–592, Geneva, May 2000.

[86] L. Lee, A. Aain, and A. Kordesch. "A 2.4-GHz CMOS tunable image-rejection low-noise amplifier with active inductor". In *Proc. IEEE Asia-Pacific Conf. Circuits Syst.*, pages 1679–1682, 2006.

[87] L. Lee, A. Aain, and A. Kordesch. "A 5-GHz CMOS tunable image-rejection low-noise amplifier". In *Proc. Int'l RF and Microwave Conf.*, pages 152–156, Putrajaya, Malaysia, Sept. 2006.

[88] H. Uyanik and N. Tarim. "Compact low voltage high-Q CMOS active inductor suitable for RF applications". *Analog Integrated Circuits and Signal Processing*, 51:191–194, 2007.

[89] F. Carreto-Castro, J. Silva-Martinez, and R. Murphy-Arteaga. "RF low-noise amplifiers in BiCMOS technologies". *IEEE Trans. Circuits and Systems II*, 46(7):974–977, Jul. 1999.

[90] A. Thanachayanont and A. Payne. "VHF CMOS integrated active inductor". *IEE Electronics Letters*, 32(11):999–1000, May 1996.

[91] K. Manetakis, S. Park, A. Payne, S. Setty, A. Thanachayanont, and C. Toumazou. "Wideband CMOS analog cells for video and wireless communications". In *Proc. Int'l Conf. Electronics, Circuits Syst.*, pages 227–230, 1996.

[92] C. Hsiao, C. Kuo, C. Ho, and Y. Chan. "Improved quality factor of 0.18μm CMOS active inductor by a feedback resistance design". *IEEE Microwave and Wireless Components Letters*, 12(2):467–469, Dec. 2002.

[93] C. Wei, H. Chiu, and W. Fend. "An ultra-wideband CMOS VCO with 3-5 GHz tuning range". In *Proc. IEEE Int'l Workshop Radio-Frequency Integration Tech.*, pages 87–90, Singapore, Nov. 2005.

[94] R. Mukhopadhyay, Y. Park, P. Sen, N. Srirattana, J. Lee, C. Lee, S. Nuttinck, A. Joseph, J. Cressler, and J. Laskar. "Reconfigurable RFICs in Si-based technologies for a compact intelligent RF front-end". *IEEE Trans. Microwave Theory and Technology*, 53(1):81–93, Jan. 2005.

[95] M. Abdalla, K. Phang, and G. Eleftheriades. "A 0.13-μm CMOS phase shifter using tunable positive/negative refractive index transmission lines". *IEEE Microwave and Wireless Components Letters*, 16(12):705–708, Dec. 2006.

[96] L. Wei, B. Ooi, Q. Xu, and P. Kooi. "High Q active inductor with loss compensation by feedback network". *IEE Electronics Letters*, 35(16):1328–1329, Aug. 1999.

[97] M. Nair, Y. Zheng, and Y. Lian. "An active inductor based low-power UWB LNA". *Proc. Int'l Conf. Ultra Wideband*, pages 813–816, Sept. 2007.

[98] A. Thanachayanont. "Low voltage CMOS fully differential active inductor and its application to RF bandpass amplifier design". In *Proc. Int'l Symp. VLSI Technology, Systems, and Applications*, pages 125–128, Apr. 2001.

[99] A. Thanachayanont. "A 1.5-V CMOS fully differential inductor-less RF bandpass amplifier". In *Proc. IEEE Int'l Symp. Circuits Syst.*, volume 1, pages 49–52, May 2001.

[100] F. Mahmoudi and C. Salama. "8 GHz 1 V CMOS quadrature down-converter for wireless applications". *Analog Integrated Circuits and Signal Processing*, 48:185–197, 2006.

[101] R. Akbari-Dilmaghani, A. Payne, and C. Toumazou. "A high Q RF CMOS differential active inductor". In *Proc. Int'l Conf. Electronics, Circuits Syst.*, volume 3, pages 157–160, Sept. 1998.

[102] M. Abdalla, G. Eleftheriades, and K. Phang. "A differential 0.13μm CMOS active inductor for high-frequency phase shifters". In *Proc. IEEE Int'l Symp. Circuits Syst.*, pages 3341–3344, 2006.

[103] R. Thomas and A. Rosa. *The analysis and design of linear circuits*. John Wiley and Sons, 1999.

[104] F. Yuan. "CMOS grator-C active transformers". *IET Proceedings, Part G - Circuits, Devices, and Systems*, 1(6):494–508, Dec. 2007.

[105] A. Tang, F. Yuan, and E. Law. "A new CMOS BPSK modulator with optimal transaction bandwidth control". In *Proc. IEEE Int'l Symp. Circuits Syst.*, pages 2550–2553, New Orleans, May 2007.

[106] J. Maneatis. "Low-jitter process-independent DLL and PLL based on self-biased techniques". *IEEE J. Solid-State Circuits*, 31(11):1723–1732, Nov. 1996.

[107] T. Soorapanth. *CMOS RF filtering at GHz frequency*. PhD. Dissertation, Stanford University, 2002.

[108] D. Li and Y. Tsividis. "Active LC filters on silicon". *IEE Proceedings, Part G - Circuits, Devices, and Systems*, 147(1):49–56, Feb. 2000.

[109] A. Mohieldin, E. Sanchez-Sinencio, and J. Silva-Martinez. "A 2.7 V, 1.8 GHz, 4th order tunable LC bandpass filter with ±0.25 dB passband ripple". In *Proc. ESSCIRC.*, pages 343–346, 2002.

[110] P. Madsen. *Q-enhancement in RF CMOS filters*. PhD. Dissertation, Aalborg University, Denmark, 2005.

[111] J. Ge. *A Q-enhanced 3.6 GHz tunable CMOS bandpass filter for wide-band wireless application*. MASc. thesis, University of Saskatchewan, 2004.

[112] W. Gee. *CMOS integrated LC Q-enhanced RF filters for wireless receivers*. PhD. Dissertation, Georgia Institute of Technology, 2005.

[113] S. Li. *RF on-chip filters using Q-enhanced LC filters*. PhD. Dissertation, Georgia Institute of Technology, 2005.

[114] A. Thanachayanont and A. Payne. "A 3-V RF CMOS bandpass amplifier using an active inductor". In *Proc. IEEE Int'l Symp. Circuits Syst.*, volume 1, pages 440–443, June 1998.

[115] A. Thanachayanont and S. Ngow. "Inductor-less RF amplifier with tunable band-selection and image rejection". In *Proc. IEEE Int'l Symp. Circuits Syst.*, volume 1, pages 573–576, May 2003.

[116] A. Sedra and K. Smith. *Microelectronics circuits*. Oxford University Press, London, 1998.

[117] F. Trofimenkoff, D. Treleaven, and L. Bruton. "Noise performance of RC-active quadratic filter sections". *IEEE Trans. on Circuit Theory*, CT-20(5):524–532, Sept. 1973.

[118] F. Trofimenkoff. "Noise margins of band-pass filters". *IEEE Trans. on Circuit Theory*, CT-20(2):171–173, Mar. 1973.

[119] H. Bachler and W. Guggenbuhl. "Noise analysis and comparison of second-order networks containing a single amplifier". *IEEE Trans. Circuit Theory*, CAS-27(2):85–91, Feb. 1980.

[120] W. Zhou, J. Pineday de Gyvez, and E. Sanchez-Sinencio. "Programmable low noise amplifier with active inductor load". In *Proc. IEEE Int'l Symp. Circuits Syst.*, volume 4, pages 365–368, May 1998.

[121] K. Sharaf. "2-V, 1-GHz CMOS inductor-less LNAs with 2-3dB NF". In *Proc. Int'l Conf. Microelectronics*, pages 379–382, Tehran, Nov. 2000.

[122] J. Yang, Y. Cheng, T. Hsu, T. Hsu, and C. Lee. "A 1.75 GHz inductor-less CMOS low noise amplifier with high-Q active inductor load". In *Proc. IEEE MidWest Symp. Circuits Syst.*, volume 2, pages 816–819, Aug. 2001.

[123] A. Pascht, J. Fischer, and M. Berroth. "A CMOS low noise amplifier at 2.4 GHz with active inductor load". In *Proc. Topical Meeting on Silicon Monolithic Integrated Circuits in RF systems*, pages 1–5, Sept. 2001.

[124] R. Weng and P. Lin. "A 2V CMOS low noise amplifier with tunable image filtering". In *Proc. IEEE Asia-Pacific Conf. Circuits Syst.*, pages 293–296, 2004.

[125] L. Lee, A. Aain, and A. Kordesch. "A 2.4 GHz CMOS tunable image-rejection low-noise amplifier with active inductor". In *Proc. IEEE Asic-Pacific Conf. Circuits Syst.*, pages 1679–1682, Dec. 2006.

[126] L. Lee, A. Aain, and A. Kordesch. "A 5 GHz CMOS tunable image-rejection low-noise amplifier". In *Proc. Int'l RF Microwave Conf.*, pages 152–156, Putrajaya, Malaysia, Sept. 2006.

[127] H. Lee, C. Lin, C. Wu, S. Liu, C. Wang, and H. Tsao. "A 15mW 69dB 2Gsamples/s CMOS analog front-end for low-band UWB applications". In *Proc. IEEE Int'l Symp. Circuits Syst.*, volume 1, pages 368–371, May 2005.

[128] M. Maadani and M. Atarodi. "A low-area, 0.18μm CMOS, 10 Gb/s optical receiver analog front end". In *Proc. IEEE Int'l Symp. Circuits Syst.*, pages 3904–3907, May 2007.

[129] W. Chen and C. Lu. "Design and analysis of a 2.5-Gbps optical receiver analog front-end in a 0.35μm digital CMOS technology". *IEEE Trans. on Circuits and Systems I.*, 53(4):977–983, Apr. 2006.

[130] C. Charles and D. Allstot. "Design consideration for integrated CMOS reflective-type phase shifters". *Analog Integrated Circuits and Signal Processing*, 50:221–229, Feb. 2007.

[131] F. Ellinger, H. Jackel, and W. Bachtold. "Varactor-loaded transmission-line phase shifter at C-band using lumped elements". *IEEE Trans. Microwave Theory and Techniques*, 51(4):1135–1140, Apr. 2003.

[132] M. Abdalla, H. Phang, and G. Eleftheriades. "A bi-directional electronically tunable CMOS phase shifter using the high-pass topology". In *Proc. IEEE Int'l Microwave Symp.*, pages 2173–2176, June 2007.

[133] J. Jiang and F. Yuan. "A new CMOS class AB transmitter for 10 Gbps serial links". *Analog Integrated Circuits and Signal Processing*, 47(3), May 2006.

[134] M. Lee, W. Dally, and P. Chiang. "Low-power area-efficient high-speed I/O circuit techniques". *IEEE J. Solid-State Circuits*, 35(11):1591–1599, Nov. 2000.

[135] F. Yuan and M. Li. "A new CMOS class AB serial link transmitter with low supply voltage sensitivity". *Analog Integrated Circuits and Signal Processing*, 49(2):171–180, Nov. 2006.

[136] X. Yang and J. Lin. "Design and analysis of a low-power discrete phase modulator in a 0.13μm logic CMOS process". *IEEE Microwave and Wireless Components Letters*, 16(3):137–139, Mar. 2006.

[137] C. Hung and T. Wai. "BPSK modulator using VCCS and resonator without carrier signal and balance modulator". *IEE Electronics Letters*, 33(15):1286–1287, Jul. 1997.

[138] B. Wupperman, B. Fox, R. Walker, S. Atkinson, D. Budin, C. Lanzi, and S. Bleiweiss. "A 16-PSK modulator with phase error correction". In *Proc. IEEE Int'l Solid-State Circuits Conf.*, pages 138–139, Feb. 1993.

[139] D. Leeson. "A simple model of feedback oscillator noise spectrum". *Proceedings of IEEE*, 54(2):329–330, Feb. 1966.

[140] T. Weigandt, B. Kim, and P. Grey. "Analysis of timing jitter in ring oscillators". In *Proc. IEEE Int'l Symp. Circuits Syst.*, pages 27–30, London 1994.

[141] T. Weigandt. *Low-phase-noise, low-timing-jitter design techniques for delay cell based VCOs and frequency synthesizer*. PhD. Dissertation, University of California, Berkeley, 1998.

[142] A. Hajimiri and T. Lee. "Design issues in CMOS differential LC oscillators". *IEEE J. Solid-State Circuits*, 34(5):717–724, May 1999.

[143] F. Yuan. "A fully differential VCO cell with active inductors for Gbps serial links". *Analog Integrated Circuits and Signal Processing*, 47(2):213–223, May 2006.

[144] J. Lee and B. Kim. "A low-noise fast-lock phase-locked loop with adaptive bandwidth control". *IEEE J. Solid-State Circuits*, 35(8):1137–1145, Aug. 2000.

[145] J. Kim, S. Lee, T. Jung, C. Kim, S. Cho, and B. Kim. "A low-jitter mixed-mode DLL for high-speed DRAM applications". *IEEE J. Solid-State Circuits*, 35(10):1430–1436, Oct. 2000.

[146] C. Park and B. Kim. "A low-noise, 900-MHz VCO in 0.6μm CMOS". *IEEE J. Solid-State Circuits*, 34(5):586–591, May 1999.

[147] Y. Eken and J. Uyemura. "A 5.9-GHz voltage-controlled ring oscillator in 0.18μm CMOS". *IEEE J. Solid-State Circuits*, 39(1):230–233, Jan. 2004.

[148] D. Ham and A. Hajimiri. "Concepts and methods in optimization of integrated LC VCOs". *IEEE J. Solid-State Circuits*, 36(6):896–909, June 2001.

[149] A. Tang, F. Yuan, and E. Law. "Low-noise CMOS active transformer voltage-controlled oscillators". In *Proc. IEEE Mid-West Symp. Circuits Syst.*, Montreal, 2007. Accepted for publication.

[150] N. Nise. *Control systems engineering*. Addison-Wesley, 2000.

[151] K. Kundert. *Predicting the phase noise and jitter of PLL-based frequency synthesizers*. Designer's Guide Consulting Inc., 2006.

[152] M. Takahashi, K. Ogawa, and K. Kundert. "VCO jitter simulation and its comparison with measurement". In *Proc. Asia and South Pacific Design Automation Conf.*, volume 1, pages 85–88, 1999.

[153] A. Demir, E. Liu, A. Sangiovanni-Vincenelli, and I. Vassiliou. "Behavioral simulation techniques for phase/delay-locked loops". In *Proc. IEEE Custom Integrated Circuit Conf.*, pages 453–456, 1994.

[154] J. Yuan and C. Svensson. "New single-clock CMOS latches and flipflops with improved speed and power savings". *IEEE J. Solid-State Circuits*, 32(1):62–69, Jan. 1997.

[155] D. DiClemente, F. Yuan, and A. Tang. "Current-mode phase-locked loops with CMOS active transformers". *IEEE Trans. on Circuits Syst. II.*, Accepted for publication in Feb. 2008.

Index

About the Author

Fei Yuan received the Ph.D. degree in electrical engineering from University of Waterloo, Canada in October 1999. He is currently an Associate Professor in the Department of Electrical and Computer Engineering, Ryerson University, Toronto, Canada. He is the author of the book *CMOS Current-Mode Circuits for Data Communications* (Springer, New York, 2006), the co-author of the book *Computer Methods for Analysis of Mixed-Mode Switching Circuits* (with Ajoy Opal, Kluwer Academic Publishers, Boston, 2004), and the author / co-author of over 120 refereed journal and conference papers in the field of mixed-mode circuits.

Dr. Yuan is the recipient of the Ryerson Research Chair award from Ryerson University in 2005, the Research Excellence award from the Faculty of Engineering, Architecture, and Science of Ryerson University in 2004, the Doctoral Scholarship from Natural Science and Engineering Research Council (NSERC) of Canada during 1997-1998, and the Teaching Excellence award from Changzhou Institute of Technology, Jiangsu, China in 1988. He has been an Adjunct Professor at University of Waterloo since 2000. Dr. Yuan is a registered professional engineer in the province of Ontario, Canada and a senior member of IEEE.

Printed in the United States
116563LV00003B/28-78/P